REACTIONS

Reactions

THE PRIVATE LIFE OF ATOMS

by **PETER ATKINS**

OXFORD
UNIVERSITY PRESS

OXFORD

UNIVERSITY PRESS

Great Clarendon Street, Oxford OX2 6DP

Oxford University Press is a department of the University of Oxford.
It furthers the University's objective of excellence in research, scholarship,
and education by publishing worldwide in

Oxford New York

Auckland Cape Town Dar es Salaam Hong Kong Karachi
Kuala Lumpur Madrid Melbourne Mexico City Nairobi
New Delhi Shanghai Taipei Toronto

With offices in

Argentina Austria Brazil Chile Czech Republic France Greece
Guatemala Hungary Italy Japan Poland Portugal Singapore
South Korea Switzerland Thailand Turkey Ukraine Vietnam

Oxford is a registered trade mark of Oxford University Press
in the UK and in certain other countries

Published in the United States
by Oxford University Press Inc., New York

British Library Cataloguing in Publication Data
Data available

Library of Congress Cataloging in Publication Data
Data available

Printed & bound in China
by C&C Offset Printing Co Ltd

ISBN 978-0-19-969512-6

1 3 5 7 9 10 8 6 4 2

CONTENTS

PREFACE

At the heart of chemistry lie reactions. When chemists shake, stir, and boil their various fluids, they are actually coaxing atoms to form new links, links that result in forms of matter that perhaps have never existed before in the universe. But what is actually going on? What form does that coaxing take? How, using the laboratory equivalents of using shovels and buckets, are individual, invisible, submicroscopic atoms urged into new partnerships?

Chemistry is thought to be an arcane subject, one from which whole populations seems to have recoiled, and one that many think can be understood only by the monkishly initiated. It is thought to be abstract because all its explanations are in terms of scarcely imaginable atoms. But, in fact, once you accept that atoms are real and imaginable as they go about their daily lives, the theatre of chemical change becomes open to visualization.

In this book I have set out to help you understand and visualize the private lives of atoms to that when you look at chemical change—and chemical change is all around and within us, from the falling of a leaf through the digestion of food to the beating of a heart and even the forming of a thought, let alone the great industrial enterprises that manufacture the modern world—you will be able to imagine what is going on at a molecular scale. In the sections that follow, I invite you to imagine constructing a toolbox of fundamental processes which will enable you to imagine levering one atom away from its partner and encouraging it to join another. Then, with those basic tools in mind, I help you to establish a workshop where you will assemble to tools and bring them to bear on a variety of projects. Finally, I introduce you, in outline but not in detail, to how those workshops are invoked to engineer certain grand projects of construction.

The representation of atoms and molecules is fraught with danger and the representation of the changes they undergo is even more hazardous. I have used drawings of molecules, cartoons really, that chemists typically used to represent their ideas, and have tried to represent various quite complicated processes in a simple and direct manner. Detail and sophistication, if you want them, can come later from other sources: I did not want them to stand in the path of this introduction and encouragement to understanding. My aim is not so much to show you exactly what is going on during a reaction but to invite you into the possibility of thinking about the private lives of atoms in a visual way, to show that chemistry is indeed all about tangible entities with characteristics that are the equivalent of personalities and which, like human personalities, lead them into a variety of combinations.

I wrote and illustrated the text myself. For reasons related to how the illustrations would lie on the page I also needed to set the pages. In that process I had a lot of help from the editorial and design departments of my publishers, who also took my necessarily somewhat amateurish raw efforts and refined them into the current version. I am very grateful to them; having gone through the entire process of constructing a book, except for its actual printing, I can appreciate even more their skills.

PWA

March, 2011;

the International Year of Chemistry

I

THE BASIC TOOLS

In this section I introduce you to the hammers, spanners, and chisels of chemistry. Here you will meet the basic types of chemical reaction that underlie all the processes around us, the processes of industry, the processes of life and death, and the processes that chemists seek to induce in their bubbling flasks. They are all the basic tools used for the fabrication of different kinds of matter.

The difference between real tools and a chemist's tools, is that the latter are exquisitely refined, for they need to shift atoms around. To make a new form of matter, perhaps one that does not exist anywhere else in the universe or simply to satisfy an existing demand, a chemist needs to be able to cajole, induce, tempt, batter, urge individual atoms to leave their current partners in one substance and join those from another substance. The new linkages must also be organized in specific ways, sometimes in assemblages of great intricacy. In this way from raw material new matter emerges. Of course, chemists do not do this atomic disassembling and re-assembling atom by individual atom: they do it by mixing, heating, and stirring their multicoloured liquids, vapours, and solids. Yet beneath these large-scale activities, the myriad atoms of their mixtures are responding one by one. Knowing what happens on the scale of atoms will help you understand what mixing, boiling, and stirring are bringing about. So, in each case I shall show you what is happening to the atoms when a common technique, a basic tool, is employed.

In later parts I shall assemble all these individual basic tools into a metaphorical chemical workshop, and then take you out to reveal construction sites of extraordinary beauty.

A Preliminary Remark

WATER AND FRIENDS

Water is the most miraculous of fluids. As well as being ubiquitous on Earth and essential for life as we know it, it has remarkable properties which at first sight don't seem to be consistent with its almost laughably simple chemical composition. Each molecule of water consists of a single oxygen atom (O) and two hydrogen atoms (H); its chemical formula is therefore, as just about everyone already knows, H_2O.

Here is one odd but hugely important anomalous property. A water molecule is only slightly heavier than a methane molecule (CH_4; C denotes a carbon atom) and an ammonia molecule (NH_3, N denotes a nitrogen atom). However, whereas methane and ammonia are gases, water is a liquid at room temperature. Water is also nearly unique in so far as its solid form, ice, is less dense than its liquid form, so ice floats on water. Icebergs float in water; methanebergs and ammoniabergs would both sink in their respective liquids in an extraterrestrial alien world, rendering their *Titanics* but not their *Nautiluses* safer than ours.

Another very important property is that water is an excellent solvent, being able to dissolve gases and many solids. One consequence of this ability is that water is a common medium for chemical reactions. Once substances are dissolved in it, their molecules can move reasonably freely, meet other dissolved substances, and react with them. As a result, water will figure large in this book and this preliminary comment is important for understanding what is to come.

The water molecule

You need to get to know the H_2O molecule intimately, for from it spring all the properties that make water so miraculous and, more prosaically, so useful. The molecule also figures frequently in the illustrations, usually looking like 1, where the red sphere denotes an O atom and the pale grey spheres represent H atoms. Actual molecules are not coloured and are not made up of discrete spheres; maybe 2 is a better depiction, but it is less informative. I shall use the latter representation only when I want to draw your attention to the way that electrons spread over the atoms and bind them together.

Each atom consists of a minute, positively charged nucleus surrounded by a cloud of negatively charged electrons. These atomic electron clouds merge and spread over the entire molecule, as in 2, and are responsible for holding the molecule together in its characteristic shape. A detail that will prove enormously significant throughout this book is that a bond between an O atom and an H atom, which is denoted O–H, consists of just two electrons. That two-electron character is a common feature of all chemical bonds.

The most important feature of an H_2O molecule for what follows is that although it is electrically neutral overall, the electrical charge is not distributed uniformly. It turns out that the O atom is slightly negatively charged and the H atoms each have a slight positive charge, 3. Throughout this book, when it is necessary to depict electric charge I shall represent positive charge by blue and negative charge by red. You need to distinguish these colour depictions from those I use to denote atoms of different elements, such as red for oxygen and blue for nitrogen! The slight negative charge of the O atom, which is called a 'partial charge', arises from an accumulation of the electron cloud on it. The electrons are drawn there by the relatively high charge of oxygen's nucleus. That accumulation is at the expense of the hydrogen atoms, with their relatively weakly charged nuclei. At their positions the cloud is depleted and the positive charge of their nuclei shines through the thinned cloud and gives them both a partial positive charge.

As a result of the attraction between opposite partial charges, one H_2O molecule can stick (loosely, not rigidly) to neighbouring H_2O molecules, and they in turn can stick to other neighbours. The mobile swarm of molecules so formed constitutes the familiar wet fluid we know as 'water'. This behaviour is in contrast to that of methane. Not only does a CH_4 molecule, 4, have much smaller partial charges because the nucleus of a C atom is more weakly charged than that of an O atom, but the partial charge of the C atom is hidden behind the surrounding four H atoms, 5. As a result, CH_4 molecules stick together only very weakly, and at room temperature methane is a gas of independent, freely moving, widely separated molecules.

FIG. 1 *Liquid water*

Liquid and solid water

Let's consider the swarm of molecules that makes up liquid water. A glance at Figure 1 shows the kind of molecular arrangement you should have in mind when thinking about the pure liquid. Think of the image of being only a single frame of a movie: the molecules are in fact in ceaseless motion, tumbling over and over and wriggling past their neighbours.

When water freezes, this motion is stilled and the molecules settle down into a highly ordered, largely stationary arrangement (Figure 2). Each molecule is still attracted to its neighbours by the attraction between opposite partial charges, but now they adopt an open honeycomb-like structure, just rocking quietly in place, not moving past one another. Melting is the collapse of this structure when the rocking motion becomes so vigorous as the temperature is raised that the molecules start to move past their neighbours and the open structure collapses. As a result of the relatively open molecular structure of ice compared to the collapsed molecular rubble of liquid water, ice is less dense than water and so can float on its own liquid.

Dissolving

I have remarked that water is a remarkably good solvent. Substances as different as salt and sugar dissolve in it readily. The oceans are great repositories of dissolved matter, including the gases that make up the atmosphere. The power of water to dissolve also springs from the presence of small electric charges on its molecules.

To understand the role of electric charges in this connection, you need to know that a substance like common salt, sodium chloride

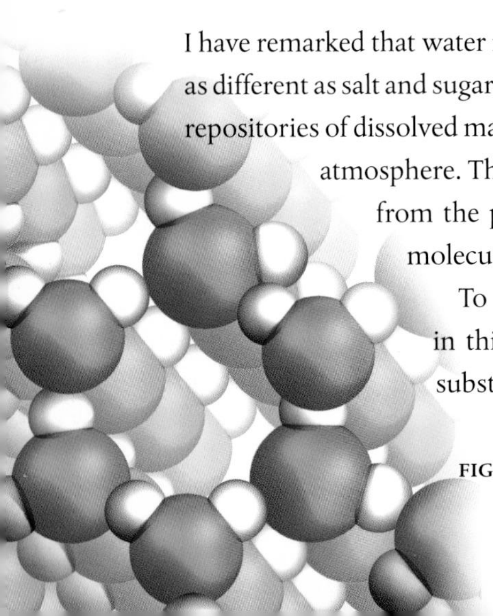

FIG. 2 *Ice*

FIG. 3 *Solid sodium chloride*

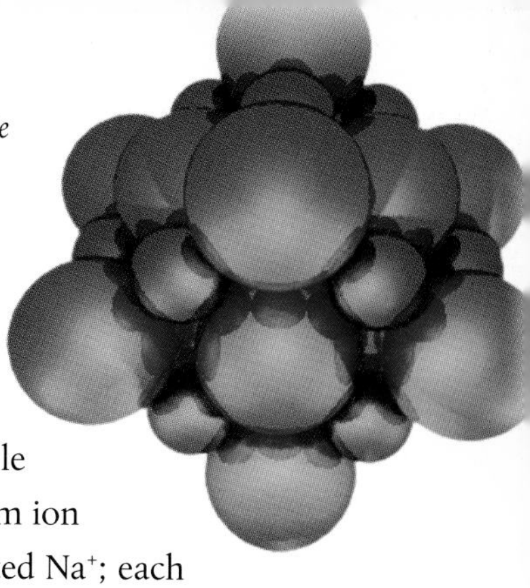

(NaCl), consists of myriad 'ions', or electrically charged atoms, stacked together in a
vast array and held together by the powerful attraction between their opposite charges
(Figure 3). Common salt is therefore an example
of an 'ionic compound'. In its case, each sodium ion
has a single positive charge (blue) and is denoted Na^+; each
chlorine ion has a single negative charge (red) and is denoted Cl^-. A
sodium ion is formed by the loss of a single electron from a sodium
atom, and a chlorine ion (more formally, a 'chloride' ion) is formed by
the acquisition of that electron by a chlorine atom. When you pick up
a grain of salt, you are picking up more ions than there are stars in the
visible universe.

Water molecules can form a 'fifth column' of subversive infiltrators
between ions and bring about the downfall of an ionic solid (Figure
4). The partial positive charges on the H atoms can simulate the full
positive charge of a sodium ion, especially when several water molecules are present, and as a result a chloride ion can be seduced into
leaving its sodium neighbours. Likewise, the partial negative charge
of each O atom of several water molecules can simulate the full negative charge of a chloride ion, and seduce a sodium ion into leaving its
chloride ion neighbours. Thus, the sodium and chloride ions can be
induced to drift off into surrounding water. Dissolution is seduction
by electrical deception.

Not all ions can be fooled by water in this way. In
some cases the electrical attraction between
neighbouring ions is just too strong to be
simulated by the relatively weak interaction of the partial charges of some H_2O
molecules. The ions remain faithful to
one another, withstand the seduction
of partial charges, and the substance

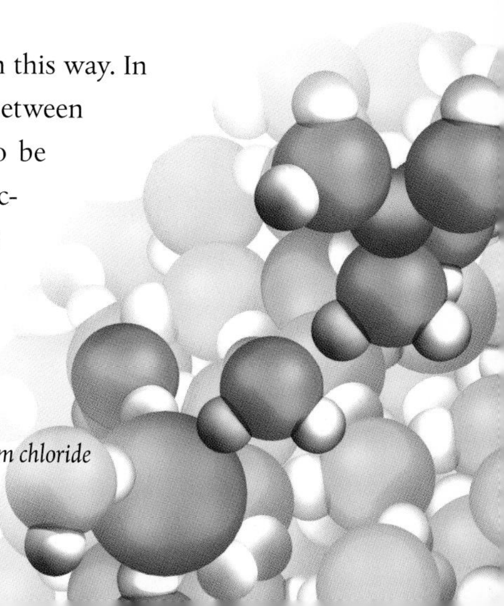

FIG. 4 *Dissolved sodium chloride*

is insoluble. This is the case with silver chloride (AgCl, Figure 5; Ag is the symbol for silver, *argentum*), an insoluble white solid. Much of our landscape survives because water is unable to dissolve the rocks. All rocks, though, are slightly soluble, and water can erode them and thereby fashion the landscape into valleys and deep canyons.

Not all compounds are ionic. Water is an example of a 'covalent compound' in which the atoms are held to one another by the electron cloud that spreads over them, as I explained above. Later in the book I shall introduce you more fully to

6

the so-called 'organic molecules', which are molecules of covalent compounds built principally but not solely from carbon. Organic molecules, which are so-called because they were once erroneously thought to be made only by living organisms, typically also contain hydrogen and commonly oxygen and nitrogen. An example is ethanol, ordinary 'alcohol', CH_3CH_2OH, **6**. Incidentally, this formula is an example of how chemists report the composition of a molecule not just by showing how many atoms of each element are present, as in C_2H_6O, but also hinting at how they are grouped together. You should compare the formula CH_3CH_2OH with the structure to identify the CH_3 group, the CH_2 group, and the OH group.

Although a lot of organic molecules do dissolve in water (think sugar), a lot don't (think oil). The difference can be traced in large measure to the fact that if atoms other than C and H are present, then the molecules have partial charges that can be emulated by water. That is the case with sugar. Glucose, for instance, is $C_6H_{12}O_6$, 7. If only C and H are present, as is the case with hydro-

7

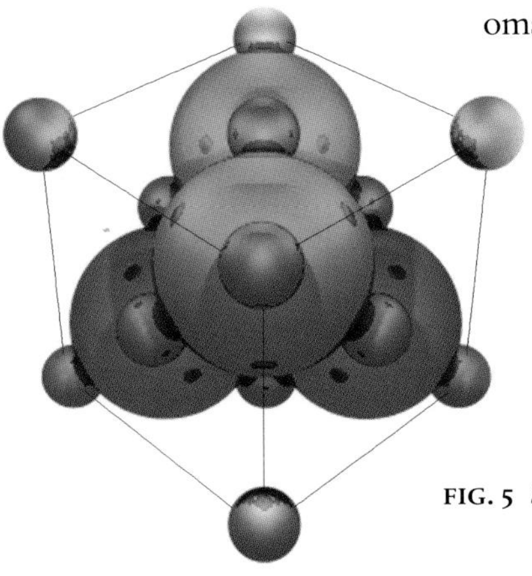

FIG. 5 *Solid silver chloride*

FIG. 6 *Alcohol (ethanol)*

carbon oils, **8**, then the partial charges are so
weak that water cannot seduce them.

Moreover, water is actually chemically
aggressive, and can react with and destroy the
compounds dissolved in it. Cooks use that characteristic to release flavours and break down cell walls.
Many organic compounds, however, do dissolve in other and
less chemically aggressive organic liquids, so many of the reactions

characteristic of organic chemistry are carried out in organic solvents such as alcohol
(Figure 6). At this stage all you need is to be
alert to that feature, and I shall expand on it
when more detail is needed.

9

1
Matter Falling Out
PRECIPITATION

I shall now introduce you to one of the simplest kinds of chemical reaction: precipitation, the falling out from solution of newly formed solid, powdery matter when two solutions are mixed together. The process is really very simple and, I have to admit, not very interesting. However, I am treating it as your first encounter with creating a different form of matter from two starting materials, so please be patient as there are much more interesting processes to come. I would like you to regard it as a warming-up exercise for thinking about and visualizing chemical reactions at a molecular level. Not much is going on, so the steps of the reaction are reasonably easy to follow.

There isn't much to do to bring about a precipitation reaction. Two soluble substances are dissolved in water, one solution is poured into the other, and—providing the starting materials are well chosen—an

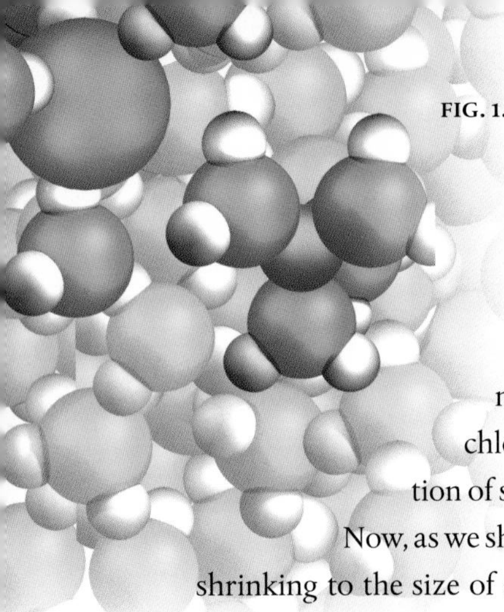

FIG. 1.1 *Sodium chloride solution*

insoluble powdery solid immediately forms and makes the solution cloudy. For instance, a white precipitate of insoluble silver chloride, looking a bit like curdled milk, is formed when a solution of sodium chloride (common salt) is poured into a solution of silver nitrate.

Now, as we shall do many times in this book, let's imagine shrinking to the size of a molecule and watch what happens when the sodium chloride solution is poured into the silver nitrate solution. As you saw in my *Preliminary remark*, when solid sodium chloride dissolves in water, Na^+ ions and Cl^- ions are seduced by water molecules into leaving the crystals of the original solid and spreading through the solution (Figure 1.1). Silver nitrate is $AgNO_3$; Ag denotes a silver atom, which is present as the positive ion Ag^+; NO_3^- is a negatively charged 'nitrate ion', 1. Silver nitrate is soluble because the negative charge of the nitrate ion is spread over its four atoms rather than concentrated on one, 2, as it is for the chloride ion, and as a result it has rather weak interactions with the neighbouring Ag^+ ions in the solid. For the same reason, the smeared out charge of the nitrate ion and its consequent weak attraction for neighbouring positively charged ions, most nitrates are soluble regardless of their accompanying positive ions. In the second solution, Ag^+ and NO_3^- ions are dispersed among the water molecules, just like in a solution of sodium chloride (Figure 1.2).

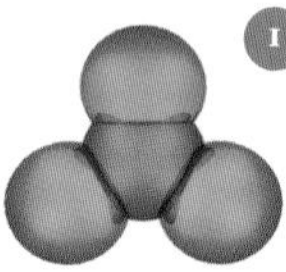

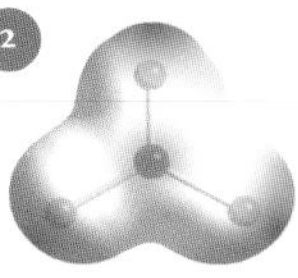

As soon as the solutions mix and the ions can mingle (Figure 1.3), the strong electrical attraction between the op-

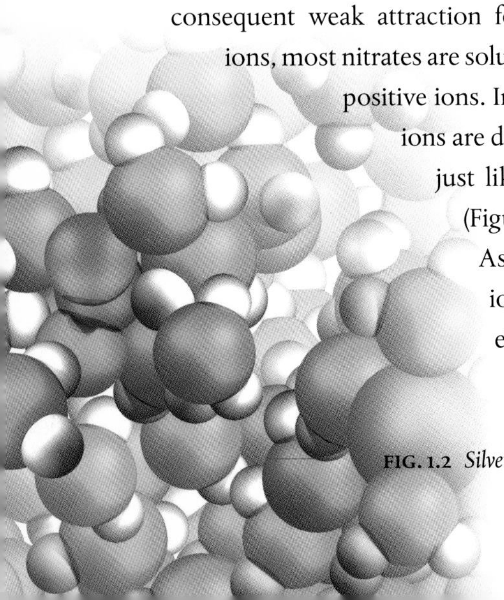

REACTIONS

12

FIG. 1.2 *Silver nitrate solution*

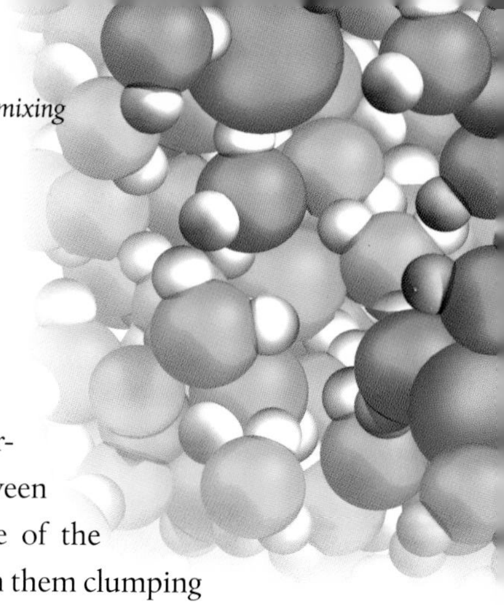

FIG. 1.3 *The solutions mixing*

positely charged Ag^+ and Cl^- ions draws them together into little localized solid clumps, a powder. To us molecule-sized observers, the tiny particles of powder are like great rocks smashing down around us, thundering down from the solution overhead (Figure 1.4). The weak interactions between the Na^+ ions and the smeared out charge of the NO_3^- ions are not strong enough to result in them clumping together: they remain in solution as a solution of soluble sodium nitrate.

Precipitation reactions are about as simple as you can get in chemistry, the chemical equivalent of wife-swapping without the moral hesitation. Nevertheless, they can be useful. Commercial examples of precipitation reactions are the preparation of silver chloride and its cousins silver bromide and silver iodide for photographic emulsions. The bright yellow pigment 'chrome yellow' is formed by a precipitation reaction in which a solution of lead nitrate (a soluble white solid) is mixed with a solution of sodium chromate, when insoluble yellow lead chromate precipitates leaving sodium nitrate in solution. On almost the very last page of this book you will see how a precipitation reaction can be used in the synthesis of a highly important drug.

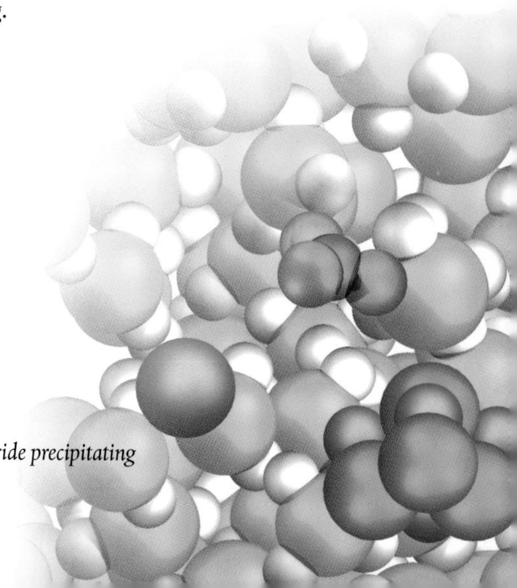

FIG. 1.4 *Silver chloride precipitating*

Give and Take

NEUTRALIZATION

The almost infinite can spring from the almost infinitesimal. Two almost infinitesimally small fundamental particles are of considerable interest to chemists: the proton and the electron. As to the almost infinite that springs from them, almost the whole of the processes that constitute what we call 'life' can be traced to the transfer of one or other of these particles from one molecule to another in a giant network of reactions going on inside our cells. I think it quite remarkable, and rather wonderful, that a hugely complex network of extremely simple processes in which protons and electrons hop from one molecule to another, sometimes dragging groups of atoms with them, sometimes not, results in our formation, our growth, and all our activities. Even thinking about proton and electron transfer, as you are now, involves them. Here I consider the transfer of a proton in some straightforward reactions in preparation for seeing later, in the second part of the book, how the same processes result in eating, growing, reproducing, and thinking. For reactions that involve the transfer of electrons, see Reaction 5.

Meet the proton

What is a proton? For physicists, a proton is a minute, positively charged, very stable cluster of three quarks; they denote it p. For chemists, who are less concerned with ultimate things, a proton is the nucleus of a hydrogen atom; they commonly denote it H^+ to signify that it is a hydrogen atom stripped of its one electron, a hydrogen ion. I shall flit between referring to this fundamental particle as a proton or a hydrogen ion as the fancy takes me: they are synonyms and the choice of name depends on convention and context.

An atom is extraordinarily small, but a proton is about 100 000 times smaller than an atom. If you were to think of an atom as being the size of a football stadium, then a proton would be the size of a fly at its centre. It is nearly 2000 times as heavy as an electron. Nevertheless, a proton is still light and nimble enough to be able to slip reasonably easily out from its home at the centre of a hydrogen atom in some types of hydrogen-containing molecules. Having escaped, it can stick to the electron clouds of certain other molecules, cloak itself with a shared pair of their electrons, and become a hydrogen atom attached to that other molecule. There, in a nutshell, is the topic of this section: proton transfer, the escape of a proton from one molecule and its capture by another. Why I have used the term 'neutralization' in the title will become clear very soon.

Physicists discovered the proton in 1919 although the concept had been lurking in their general awareness ever since Ernest Rutherford (1871–1937) had shown in 1911 that an atom was mostly empty space with a central core, the nucleus. The structure of an atomic nucleus soon became clear: it was found to consist of a certain number of protons and the proton's electrically neutral cousin, the neutron. By 1913 Henry Moseley (1887–1915, shot at Gallipoli by protons bundled together as iron nuclei) had determined the numbers of protons in the nuclei of the atoms of many elements. Thus, a hydrogen nucleus is a single proton, there are two protons in the nucleus of helium, three in lithium, 26 in iron, and so on.

Chemists brought protons fully into their vocabulary in 1923 but had unwittingly been shuttling them around between molecules of various kinds, thinking of them as 'hydrogen ions', since the nineteenth century. Artisans and cooks had been shuttling them around, even more unwittingly, for centuries.

A little light language

I need to step back a few years to put the proton into a chemical context for you. As I remarked in the preface, chemists are always on the lookout for patterns, both patterns in the properties of the elements and patterns in the reactions that substances undergo. It had long been familiar to them and to their predecessors the alchemists that certain compounds react together in a similar way. Two of these groups of compounds that reacted together in a certain pattern came to be known as 'acids' and 'alkalis'. Because this reaction seemed to quench the acidity or alkalinity of the participants, it came to be known as 'neutralization'.

Chemists also noted that the product of a neutralization reaction between an acid and an alkali is a salt and water. A 'salt' takes its name from common salt (sodium chloride) but might be composed of other elements. Chemists often take the name of a single exemplar and use it to refer to an entire class of similar entities. A salt is an ionic compound, like sodium chloride (recall Figure 3 of my *Preliminary remark*), that is neither an acid nor an alkali.

Let's focus initially on acids and alkalis. The name 'acid' is derived from the Latin for 'sour, sharp taste', as for vinegar and lemon juice, both of which contain acids. Taste is an extraordinarily dangerous test for an acid: for some people and some acids, it would work only once! The name 'alkali' is derived from the Arabic words for 'ash', because a common source of an alkali was wood ash, a complex, impure mixture of potassium oxides, hydroxides, carbonates, and nitrates. Wood ash was heated with animal fats to produce soap in a reaction that we explore later (Reaction 18). Indeed, this is the basis of an early and particularly dangerous test for alkalis: they had a soapy feel. That they felt soapy was due to the formation of soap-like substances from the fats in the incautiously probing fingers.

The term 'alkali' has been largely superseded in chemical conversations by the more general term 'base', and I shall gradually move towards using that name. An alkali is simply a water-soluble base; there are bases that don't dissolve in water, so 'base' is a more general term than 'alkali'. The name stems from the fact that a single compound, the base, can be used as a foundation for building a series of different salts by reaction with a choice of acids. Thus, suppose you take the base sodium hydroxide, then you would get the salt sodium chloride if you neutralized it with hydrochloric acid, the salt sodium sulfate if you used sulfuric acid, and so on.

At this point I have introduced you to the terms 'acid', 'base', and 'alkali' if the base is soluble in water. The reaction between them is 'neutralization' and the product is a 'salt' and water. What, though, is an acid, and what is a base? And how can we identify them without killing ourselves in the process?

A suggestion from Sweden

The Swedish chemist Svante Arrhenius (1859–1927) took an early fruitful step when he suggested that an acid is any compound containing hydrogen that, when it dissolves in water, releases hydrogen ions.

Thus, when the gas hydrogen chloride, which consists of HCl molecules, 1, dissolves in water each molecule releases a proton from inside the hydrogen atom. Once the proton has gone, the electron cloud that spread over the proton like a wart on the side of the chlorine atom, Cl, snaps back entirely on to the Cl atom to form a chloride ion, Cl⁻. The resulting solution of H⁺ ions and Cl⁻ ions is 'hydrochloric acid'.

Much the same happens when the organic compound acetic acid, CH_3COOH, 2, the tart component of vinegar, dissolves in water. Once the molecule is surrounded by water molecules, a proton at the centre of the H atom attached to an O atom slips out of its electron cloud as an H⁺ ion. That cloud, no longer held in place by the proton, snaps back on to the O atom, forming an acetate ion, $CH_3CO_2^-$, 3.

REACTIONS

> *Pedant's point.* Although just about every HCl molecule gives up its proton, only about 1 in 10 000 acetic acid molecules gives up a proton in this way.

What about bases? Arrhenius went on to suggest that a base is a compound that, when it dissolves in water, results in the formation of hydroxide ions, OH⁻, 4. Thus, according to this view, sodium hydroxide, NaOH, is a base because when it dissolves, the sodium ions and hydroxide ions that are already present in the solid separate to give a solution of Na⁺ and OH⁻ ions.

These suggestions account for the neutralization pattern. When hydrochloric acid, which according to Arrhenius consists of dissolved H⁺ and Cl⁻ ions, is poured into a solution of sodium hydroxide, which consists of dissolved Na⁺ and OH⁻ ions, the H⁺ and OH⁻ ions immediately clump together in pairs and form a bond to give water, H–OH, which we recognize as H_2O. That removal of H^+ and OH^- ions from the solution leaves a solution of Na^+ and Cl^- ions, which jointly make up the salt sodium chloride, NaCl. Much the same happens when acetic acid is poured into sodium hydroxide solution: the H^+ ions present in the acid clump on to the OH^- ions present in the alkali, form water, and leave sodium ions and acetate ions in solution, corresponding to the salt sodium acetate.

Even compounds that don't already have OH^- ions present initially can give rise to them when they dissolve in water. For instance, when ammonia, NH_3, **5**, dissolves in water some of the molecules suck out a proton from a neighbouring H_2O molecule, become 'ammonium ions', NH_4^+, **6,** and thereby convert the water molecule into an OH^- ion. The solution now acts as an alkali by virtue of the OH^- ions it contains. When hydrochloric acid is poured into it, the H^+ and OH^- ions snap together to form water in the usual way, leaving NH_4^+ and Cl^- ions. These ions jointly form the salt ammonium chloride, NH_4Cl. Arrhenius certainly seems to have got to the heart of the pattern of neutralization.

Another suggestion from further south

Despite Arrhenius's considerable success, his conceptual butterfly net didn't capture everything that looked like a neutralization reaction. That became clear once chemists turned their attention to reactions taking place in liquids other than water and even in the absence of any solvent at all. They found that many compounds act like acids and bases even though there is no water present, yet the Arrhenius definitions involve water explicitly.

This is where the proton, H^+, comes into its own and moves to centre stage and will appear on numerous occasions throughout

FIG. 2.1 *The formation of hydrochloric acid*

this book. Almost simultaneously, in 1923, Thomas Lowry (1874–1936) in England and Johannes Brønsted (1879–1947) in Denmark suggested that all neutralization reactions are captured by a very simple idea. They suggested that an acid is anything that can donate a proton to another molecule. A base, they suggested, is anything that can accept that donated proton. In other words, an acid is a proton donor and a base is a proton acceptor. Acids are therefore compounds that have a sufficiently loose proton inside one of their hydrogen atoms for it to be able to slip away and migrate to another molecule or ion. Bases are substances that have sufficiently dense regions of electron cloud to which an incoming proton can attach. According to this view, in a neutralization reaction a proton leaves its supplier, an acid, and ends up attached to a proton acceptor, a base. In short, neutralization is proton transfer. This is the molecular give and take, the donation and accepting, of the title.

Let's see how this works. As you have seen, and as I have illustrated in Figure 2.1, if we were to watch a hydrogen chloride molecule, HCl, plunging from the gas and splashing down into water, we would see

it release a proton. The released proton doesn't just hang around unattached: it is donated to a nearby H_2O molecule, which becomes a 'hydronium ion', H_3O^+, 7, and then that ion wriggles off out of sight through the solution.

Similarly, if we shrink, imagine ourselves immersed in water, and watch pure acetic acid being mixed into the water we see that a few CH_3COOH molecules donate a proton to the neighbouring water molecules. We conclude that acetic acid, seen to be a proton donor, is indeed an acid. The three H atoms attached to the C atom in acetic acid are too tightly held to be able to escape from the grip of their surrounding electrons, so the acid character of CH_3COOH

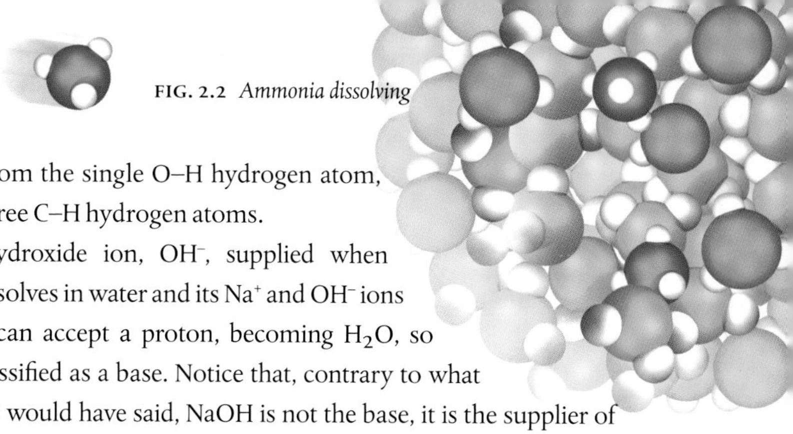

 FIG. 2.2 *Ammonia dissolving*

springs from the single O–H hydrogen atom, not the three C–H hydrogen atoms.

The hydroxide ion, OH⁻, supplied when NaOH dissolves in water and its Na⁺ and OH⁻ ions separate, can accept a proton, becoming H_2O, so OH⁻ is classified as a base. Notice that, contrary to what Arrhenius would have said, NaOH is not the base, it is the supplier of the base: the base is the proton-accepting OH⁻ ion that NaOH provides.

Figure 2.2 shows what we would see when we shrink and watch ammonia dissolve in water. After splashdown we see an NH_3 molecule accept a proton from a neighbouring water molecule and become the ammonium ion, NH_4^+. That ion then wriggles off through the surrounding water molecules and away from the OH⁻ left as a result of the proton transfer from H_2O. We conclude that because it accepts a proton, NH_3 is a base.

The consequent capture of strange fish

When definitions are enlarged, like changing from fishing in coastal water to deep ocean, peculiar species are sometimes caught. Before we go on to see that the new definition captures everything that Arrhenius would regard as an acid (and then more), there is a very important, completely unexpected fish brought up in Lowry and Brønsted's joint net.

One of the molecules with regions where the electron cloud is dense and there is enough partial negative charge for a proton to be able to attach is H_2O itself. I have already let this property slip into the discussion without comment when I remarked that proton transfer to H_2O results in the formation of a hydronium ion, H_3O^+. Now, though, we have to bite the bullet and accept that, if we go along with everything so far, then because H_2O accepts a proton, water itself is a base.

Water is a molecular fish with yet another trick up its remarkable sleeves. We have also seen that when ammonia dissolves in water, an H_2O molecule surrenders a proton to an NH_3 molecule and itself becomes OH^-. Here is a second bullet to bite: because H_2O can act as a proton donor, you now have to accept that it is also an acid!

But here is a funnier thing still. Because two-faced water is not only an acid but also a base, then even before a conventional acid or base is added to a beaker of water, the molecules already present are both acids and bases. You now have to accept that when you drink a glass of water, you are drinking an acid. This is not a trivial conclusion to be shrugged off by saying that somehow or other there probably isn't much acid present. Every molecule is an acid, so you are drinking pure, highly concentrated acid. If you don't like that thought, then you might like it even less to realise that you are also drinking a base. Once again, you can't shrug off the thought by saying that the water is probably just a very dilute solution of a base. Every molecule is a base, so with every sip or gulp you are drinking highly concentrated, pure base. Such are the consequences of expanding and generalizing definitions: designed to catch sardines, they turn out to capture sharks.

With this insight into the Janus nature of water in mind, we shrink to the size of a molecule and jointly watch what is going on in a glass of pure, dangerous water. We see one H_2O molecule acting as a proton donor, an acid, and catch sight of another H_2O molecule in the act of another accepting a proton and so acting as a base (Figure 2.3). The accepting molecule becomes a hydronium ion, H_3O^+, which we see drift away, almost certainly to donate its extra proton to another water molecule somewhere else in the liquid. When it does, it reverts to H_2O and the acceptor molecule takes up the burden and drifts off as H_3O^+. Similarly, we see the OH^- ion left after the first donation

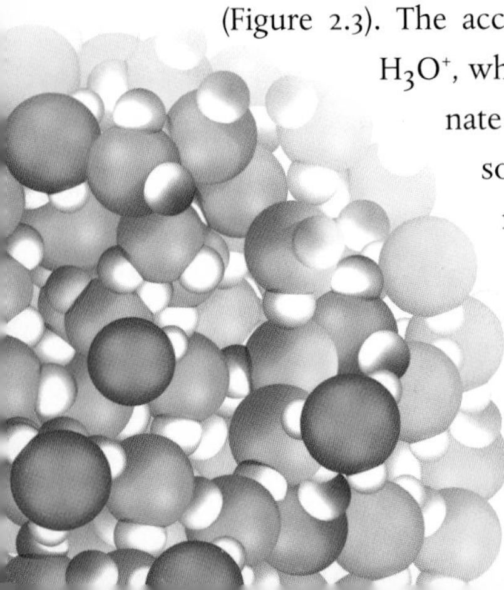

FIG. 2.3 *Water donating to itself*

REACTIONS

accept a proton from another water molecule, so becoming H_2O again, with the second water molecule taking up the baton of being OH^-, and so on.

The important point about this discussion is that pure water is by no means purely H_2O. It is overwhelmingly H_2O molecules, but immersed in it there is a scattering of OH^- ions that have been formed by proton loss and a matching number of H_3O^+ ions that have been formed by proton gain. As we stand there watching, we see protons ceaselessly being handed between molecules like hot potatoes with H_3O^+ and OH^- ions flickering briefly into existence and then very quickly reverting to H_2O again. The concentration of these ions in pure water is very low, but they are there. To get some idea of their abundance, if every letter in a 1000 page book represented an H_2O molecule, you would have to search through 10 such books to find one H_3O^+ ion or one OH^- ion. Nevertheless, the 'flexibility' of water—its dynamic nature, in the sense that there are ions present even in the pure liquid, albeit at a very low level, with protons hopping from molecule to molecule—is a crucial feature of this extraordinary liquid. It adds to the mental picture of what you should imagine when you look at a glass of water and think about its nature and, in due course, the reactions taking place there.

Finally, at last, down to business

Now I can lead you to the point of visualizing what happens at a molecular level in a neutralization reaction. Let's imagine ourselves shrunken as usual and standing together in a solution of sodium hydroxide. We see a dense forest of water molecules, and dotted here and there are sodium and hydroxide ions. Then hydrochloric acid rains in, bringing a torrent of water molecules and among them hydronium ions and chloride ions. The H_3O^+ ions in the torrent move through the solution and soon, almost instantaneously, encounter

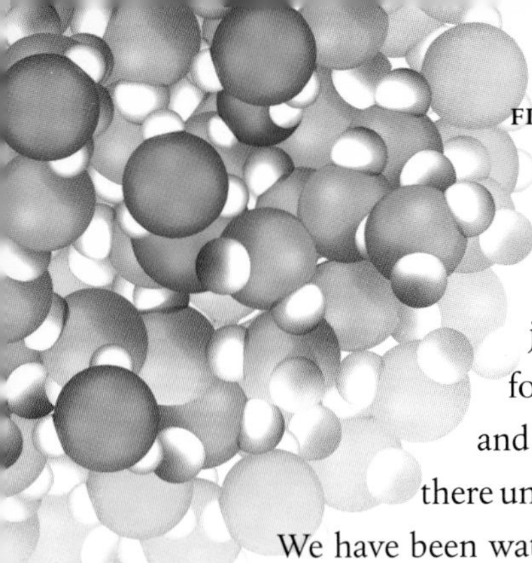

FIG. 2.4 *Neutralization*

one of the OH⁻ ions provided by the sodium hydroxide. As soon as they meet, a proton jumps across from the H_3O^+ ion to the OH⁻ ion, forming two H_2O molecules. The chloride ions and sodium ions also present in solution remain there unchanged (Figure 2.4).

We have been watching the event common to all neutralization reactions in water: a proton transfers from a hydronium ion to a hydroxide ion to form water. The salt, so characteristic of early visions of neutralization reactions is there like us only as a spectator: the real business of the reaction is proton transfer.

I remarked earlier that Arrhenius's vision was too limited because his view of acids, bases, and neutralization reactions depended on the presence of water. This restriction is removed in the proton transfer vision of neutralization reactions, as a proton can hop directly from an acid to a base without a solvent needing to be present.

To appreciate the last point, let's imagine floating in a gas of ammonia, where we are surrounded by NH_3 molecules zooming around and colliding with one another. Now someone squirts in a puff of hydrogen chloride gas with HCl molecules also zooming around and colliding with one another. When the gases mingle, collisions occur between HCl and NH_3 (Figure 2.5). The electron cloud of an NH_3 molecule is concentrated on the N atom and acts there as a sticky patch to which a proton can attach. As we watch we see that in a collision the proton of HCl sticks to that patch on NH_3, so forming NH_4^+. When the Cl⁻ ion ricochets away, it leaves the proton behind. In this way, by direct collisions, the original gas of HCl and NH_3 molecules quickly turns into a swarm of NH_4^+ and Cl⁻ ions. These ions are attracted to each other by their opposite charges and immediately clump together to form a fine white fog of solid ammonium chloride, NH_4Cl. Neutralization, proton

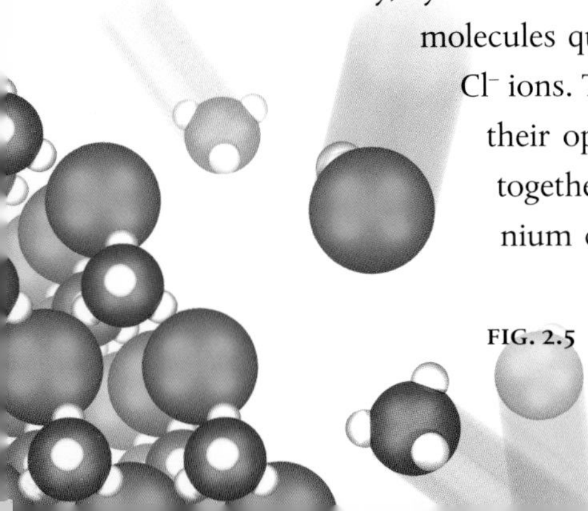

FIG. 2.5 *Proton transfer in a gas*

transfer, has occurred in the absence of water, indeed of any solvent at all.

Neutralization reactions are used to form salts when more economical sources are not available: chemists just choose solutions of the appropriate acid and base and mix them together in the right proportions. They are also used for more technical tasks, such as analyzing solutions for their content. However, as I indicated in the introduction to this reaction, proton transfer comes into its own when we turn to the reactions of life. I take up that story in Parts 2 and 3.

3
Burns Night
COMBUSTION

urning, more formally combustion, denotes burning in oxygen and more commonly in air (which is 20 per cent oxygen). Combustion is a special case of a more general term, 'oxidation', which originally meant reaction with oxygen, not necessarily accompanied by a flame. The rusting of iron is also an oxidation, but we don't normally think of it as a combustion because no flame is involved. Oxidation now has a much broader meaning than reaction with oxygen, as I shall unfold in Reaction 5. For now, I shall stick to combustion itself.

To achieve combustion, we take a fuel, which might be the methane, CH_4, **1**, of natural gas or one of the heavier hydrocar-

bons, such as octane, C_8H_{18}, **2**, that we use in internal combustion engines, mix it with air, and ignite it. The outcome of the complete combustion of any hydrocarbon is carbon dioxide and water but incomplete combustion can result in carbon monoxide and various fragments of the original hydrocarbon molecule. All combustions are 'exothermic', meaning that they release a lot of energy as heat into the surroundings. We use that energy for warmth or for driving machinery.

Another example of an exothermic combustion is provided by the metal magnesium, which gives an intense white light as well as heat when it burns in air. A part of the vigour of this reaction is due to the fact that magnesium reacts not only with oxygen but also with nitrogen, the major component of air. You should be getting a glimpse of the broader significance of the term 'oxidation' in the sense that the reaction need not involve oxygen; in magnesium's case, nitrogen can replace oxygen in the reaction. Magnesium foil was used in old-fashioned photographic flashes and in fireworks. The latter now mostly use finely powdered aluminium, which is much cheaper than magnesium and reacts in much the same way. In what follows you could easily replace aluminium with magnesium if you want to think fireworks.

For the whole of the following discussion you need to be familiar with oxygen, O_2, 3, a peculiar molecule in several respects. The two O atoms in O_2 are strapped together by a reasonably strong bond. However, the electrons responsible for the bonding are arranged in such a way—think of there being wispy gaps in an otherwise smooth cloud—that it is quite easy for other electrons to insert themselves. When electrons enter and fill the gaps, they force the molecule to fall apart, perhaps forming two O^{2-} ions. One of the gaps might also accept an electron still attached to a proton, that is, a hydrogen atom, H, resulting in the formation of OOH, 4. These unusual species will soon move onto our stage and act out their roles in combustion.

Blazing metal

The combustion of magnesium is a bit easier to talk about than the combustion of a hydrocarbon fuel, so I shall deal with magnesium first and then move on to the more familiar reaction of the combustion of hydrocarbons. I'll pretend that in its combustion, magnesium combines only with oxygen: its additional reaction with nitrogen when it burns in air doesn't add much new and complicates the discussion.

I have to admit, in addition to that simplification, that my account of the sequence of events that occurs at the vigorously changing tumultuous surface of burning magnesium is largely speculative. I am sure that you can appreciate that it is very hard to venture into the incandescent eye of the storm and make careful observations there. In fact, 'sequence of events' also gives the wrong impression of an orderly series of changes. The burning surface is at the eye of a thermal storm, with atoms being ripped off the metal in a maelstrom of processes occurring in no particular order. I will do my best to convey the essential features of what is going on, but think of my account as a series of snapshots taken more or less at random during an all-out battle.

As well as being familiar with oxygen, for this part of this discussion you also need to be familiar with one feature of magnesium. A magnesium atom has a nucleus with a fairly feeble positive charge, so the atom's outer electrons are not held very tightly. It turns out that a magnesium atom, Mg, can lose up to two electrons fairly readily, and as a result be changed into a doubly charged magnesium ion, Mg^{2+}, 5.

The combination of an oxygen molecule having the ability to sponge up electrons and a magnesium atom having only feeble parental control over its own electrons, means that oxygen molecules can accept electrons from the

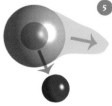

REACTIONS

FIG.3.1 *Magnesium burning*

atoms near the surface of a strip of magnesium. The illustration in Figure 3.1 is my attempt to convey the essence of what is going on. We see that atoms are being ripped out of the solid as ions where an O_2 molecule strikes the surface. The solid melts in the heat of reaction. That is, the atoms jiggle around so vigorously that they can move past one another and behave like a tiny puddle of liquid. This mobility enables the atoms to be ripped out more easily. As we watch, the Mg^{2+} and O^{2-} ions that form in the turmoil, stack together as—on a molecular scale—great rocks of the ionic compound magnesium oxide, MgO, that fly through the air and are blasted off by the currents of molecules of air. To an outside observer, this stacking together results in the formation of tiny particles of magnesium oxide, which fly off as 'smoke'.

An old flame

Rather more calmly, and proceeding by an entirely different mechanism, is the combustion of methane, such as occurs when natural gas burns. This combustion occurs in a sequence of steps that involves radicals.

I need to introduce you to radicals. A 'radical' (the old name, 'free radicals', is still widely used) is an atom or group of atoms that can be regarded as being broken off a molecule. An example is the methyl radical, $\cdot CH_3$, which is formed when the pair of electrons that makes up the carbon–carbon bond in ethane, CH_3CH_3, 6, is torn apart. In this case, the two electrons of the C–C bond are separated, and each resulting $\cdot CH_3$ radical, 7, carries away one of them, as indicated by the dot. This description is the basis of a more formal definition of a radical as a species with a single unpaired electron.

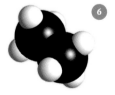

Other examples are the hydroxyl radical, ·OH, which is formed when an H–OH bond in water is broken, and a chlorine radical, ·Cl, in this case a single atom, formed when a chlorine molecule, Cl_2, is torn apart. Because they have an unpaired electron, with its hunger to pair with another unpaired electron and form a bond, most radicals are highly reactive, and do not survive for long. See Reaction 12 for a more complete discussion of radicals and their reactions.

Now that you know what a radical is I am ready to show you what happens when methane, CH_4, is ignited. Much the same happens when you ignite bottled gas, propane, $CH_3CH_2CH_3$, and even the heavier hydrocarbons of gasoline and diesel, but I shall keep it simple by focusing on methane with its single C atom.

We shall imagine ourselves shrunk and standing in a jet of natural gas, surrounded by methane and air molecules hurtling to and fro around us. We see a spark or match flame (both very interesting hot radical-rich environments in themselves!) brought up to where we are standing. It provides enough energy to break one of the C–H bonds in a methane molecule: a liberated hydrogen atom springs away and a methyl radical, ·CH_3, is formed. In this context a hydrogen atom is treated as a radical and written ·H. We now see these radicals going on the attack. As usual in a combustion reaction, there is no strict sequence of events, so think of the following remarks as trying to capture the overall turmoil going on in the battleground of the flame, not an orderly progression of snipings.

Close to us we see a hydrogen atom collide with a CH_4 molecule and pluck off one of that molecule's H atoms, so forming H_2 and leaving ·CH_3 (Figure 3.2). Elsewhere nearby we see a hydrogen atom sticking to an O_2 molecule to form HO_2·. Remember that the wispy gaps in the electron cloud of the O_2 molecule can accommodate an electron, and in particular the unpaired electron carried by a radical. As we watch we see that radical collide

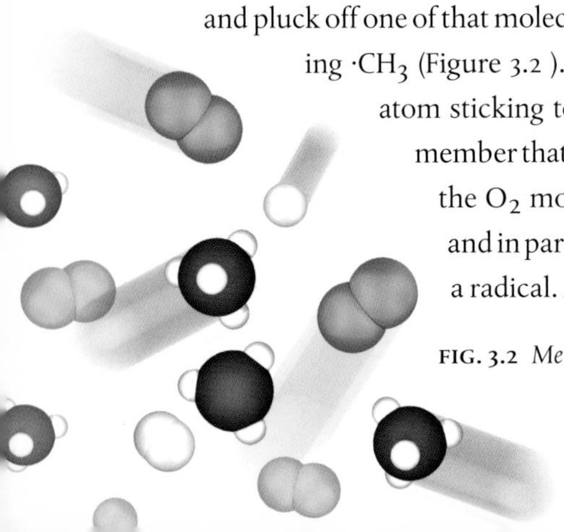

FIG. 3.2 *Methane burning*

FIG. 3.3 *Flaming methane*

with and attack another CH_4 molecule, pluck off an H atom, become HOOH, and immediately fall apart as two ·OH radicals. These virulent little radicals now join in the fray, and we see one pluck an H atom off a ·CH_3 radical to form H_2O and ·CH_2·, a two-fanged 'biradical' (Figure 3.3). As we watch we see CH_4 being whittled down to naked C as its H atoms are stripped away by radical attack. But if we look elsewhere we see a ·CH_3 radical colliding with an O_2 molecule, attach to it, and then shrug off an H_2O molecule. That little skirmish leaves ·CHO, and we realise that it is carbon on its way to becoming carbon dioxide, CO_2. Although it has moved out of sight, the H_2 formed earlier in the storm of reactions is destined for a short life, because it too comes under attack, perhaps by O_2 to form HOOH, which is hydrogen on its way to becoming H_2O.

Colourful incandescence

The battle of radicals generates both heat and light. One question that might already have occurred to you is why both natural gas and propane burn with a blue flame if there is plenty of air but with a smoky yellow flame if the supply of air is restricted.

In the tumult of a flame, with methane molecules torn apart and H atoms stripped off C atoms, there is a good chance that C atoms will collide or that fragments of methane molecules will meet, bond together by sharing their unpaired electrons, but then have their H atoms plucked off by the aggressive ·O· atoms or ·OH radicals, leaving diatomic C_2 molecules. But these will not be ordinary C_2 molecules; they will have their electron distributions distorted by the vigour of their formation, **8**. These distorted distributions immediately collapse back into the form characteristic of an ordinary C_2 molecule. The shock of that collapse generates

FIG. 3.4 *Soot formation*

a photon, typically of blue light, which carries away the excess energy. Thus, we see the flame glow blue.

In a restricted supply of air, carbon atoms have time to aggregate as C_3, C_4, and up to hundreds and even thousands of C atoms. In other words, in the absence of the enemy oxygen, small particles of solid carbon will form (Figure 3.4) , not individual molecules. Like solids in general, these particles become incandescent at the high temperatures characteristic of flames, and glow first red and then yellow hot. Thus we see a blue flame give way to a yellow flame as the oxygen is depleted, and some smoke—in this case, of a stream of carbon particles we know as soot—will form.

Sturm und drang

Under certain circumstances, a mixture of fuel and air does not burn smoothly but explodes. Explosions are fundamentally very fast reactions that generate a lot of gaseous product which immediately expands as a destructive shockwave. To understand how docile combustion can become violent explosion you need to know some more about radical reactions, and I take the story further in Reaction 12, to which you could now cautiously turn.

REACTIONS

Back to Basics
REDUCTION

I n its original meaning, reduction was what was done to a metal ore to obtain the metal itself: the stony ore hacked from the land was reduced to the malleable, ductile, lustrous, useful metal. Ores are commonly oxides or sulfides, so the process of reduction typically involves the removal of oxygen or sulfur. In that sense, reduction is the opposite of oxidation, which I touched on in Reaction 3. In this chapter I shall stick with the metallurgical context and examine that hugely important industrial process, the reduction of iron ore to iron at what can be regarded as the head of the steel chain. However, like oxidation, the concept of reduction has acquired a much broader meaning, as I shall touch on fleetingly in this section and reveal fully in Reaction 5.

A typical iron ore is *haematite*, an iron oxide of composition Fe_2O_3 and consisting of a stack of Fe^{3+} and O^{2-} ions (Figure 4.1, over the page; Fe is the symbol for iron, from the Latin *ferrum*). In the industrial

FIG. 4.1 *Haematite (an iron ore)*

process for the production of iron, the 'reducing agent', the substance that brings about reduction, is essentially carbon in the form of coke. Early furnaces used charcoal, but coke is much harder and allows for much taller columns of ore, carbon, and limestone (the last to collect impurities as slag; see Reaction 9). Reduction on this huge scale is carried out in the great blast furnaces that epitomize heavy industry and the industrial revolution, but those furnaces are little more than sophistications of the fires that first led mankind from the Bronze Age to the Iron Age about 3000 years ago.

REACTIONS

Pedant's point. I wrote the weasel phrase 'essentially carbon' because carbon is not the actual agent: it is used to make carbon monoxide, CO, which is commonly held to be the actual reducing agent.

The eponymous blast of a blast furnace is a blast of air. It may seem odd to use oxygen-rich air in a process designed to remove oxygen from an ore, but it is used to oxidize the carbon to carbon monoxide and also to help ensure that the contents of the furnace do not settle to the bottom. Care must be taken, of course, to ensure that the blast is not so strong that the contents are blown out through the top! A gas such as carbon monoxide is much more fleet of molecular foot than lumps of coke and can penetrate into and attack the molten ore wherever it lies. The partial combustion of the coke—for that is what the formation of CO represents—also serves to raise the temperature within the furnace and render the ore molten.

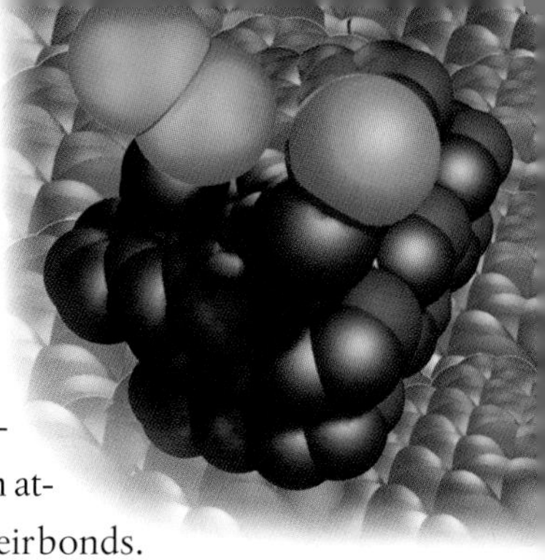

FIG. 4.2 *Coke burning*

The heart of brightness

I shall now take you in imagination into the heart and heat of the furnace. There we see the hot glowing solid coke, with its vigorously vibrating carbon atoms compressing and stretching and straining at their bonds. Whole sheets of carbon atoms on the surface of the lumps of coke are heaving in waves and breaking up under the stress of motion. Oxygen molecules are hitting the surface fragments all around us, snipping off C atoms, flying off as CO molecules, and releasing energy as strong carbon–oxygen bonds are formed (Figure 4.2).

We struggle up to another part of the flaming fiery furnace where we see the reaction of the newly formed gaseous CO with the now molten ore. As I have remarked, that ore, Fe_2O_3, is composed of iron ions, Fe^{3+}, and oxide ions, O^{2-}, but the heat has overcome the strong attraction between the small, highly charged ions and it is now a molten, mobile but viscous liquid and no longer a rigid rock. We have to see inwardly in our imagination, because the incandescent dense fluid is impossible to penetrate visually. But in that imagination we see a vigorously vibrating CO molecule home in on and attack an O^{2-} ion in the molten ore and suck out the O atom from it and become CO_2 (Figure 4.3). That stable little molecule scuttles off through the mix and in due course adds to the burden of carbon dioxide in the atmosphere. This last step, the prising out of an O atom by CO, is an example of a type of reaction I discuss in more detail later (Reaction 9).

The removal of the O atom from the O^{2-} ion leaves behind two electrons that need a home. We can imagine them flooding onto the Fe^{3+} ions that still cluster round the O^{2-} ions, even in the fluid (Figure 4.4 on the next page). It is easy to do the arithmetic: for every three O atoms extracted by CO molecules

FIG. 4.3 *Carbon monoxide extracting oxygen*

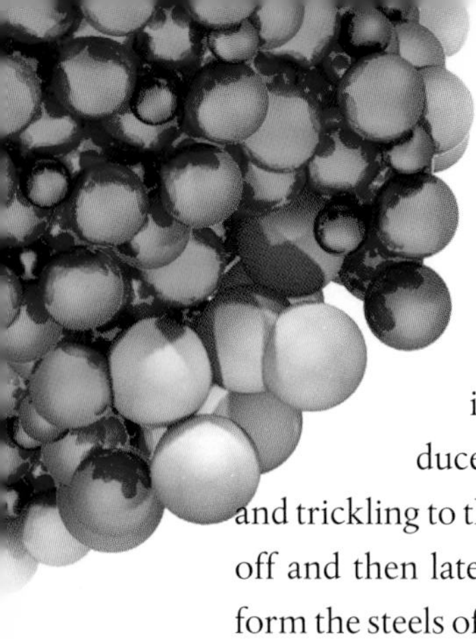

FIG. 4.4 *Iron forming*

from the O^{2-} ions in the molten ore, six electrons are discarded and are immediately picked up by two Fe^{3+} ions, so converting them to two neutral iron atoms, Fe. As we watch we see the Fe atoms produced in this way congregating to form a mobile liquid and trickling to the base of the furnace. There the liquid iron is drawn off and then later refined and alloyed with various other metals to form the steels of the world.

One fate for the iron recovered with such a seriously huge input of energy, other than its ennoblement and deployment as steel, is its sad reversion to a state similar to its original ore. I describe the death of metal in the section on corrosion, Reaction 8.

Two Hands Clapping
REDOX REACTIONS

I promised in Reactions 3 and 4 to lead you to the promised land of the modern understanding of oxidation and reduction reactions. This is the section where these two great chemical rivers flow together and acquire great explanatory power and wide applicability. I have already shown that one great class of reactions, those between acids and bases (Reaction 2), takes place by the transfer of one fundamental particle, the proton. I shall now show you that oxidation and reduction reactions all take place by the transfer of another fundamental particle, in this case the proton's cousin, the electron.

Don't be put off by the thought that in this unification of two great rivers I am embarking on a highly abstract, distant-from-reality account. All I am doing is looking for and presenting the essential step that is involved in these reactions. This is a bit like looking for the core idea of many sports, which is to get a projectile to move into a particular location, be it soccer, baseball shooting, darts, archery, or billiards.

I hope you will begin to appreciate in the course of this chapter that when chemists carry out their reactions by stirring, boiling, and mixing, all they are doing is encouraging fundamental particles, in this case electrons but in Reaction 2 protons, to migrate from where they are found to where the chemist wants them to be. Industry does the same coaxing on a massive scale.

My aim here is to show you that everything I discussed in Reactions 3 and 4 boils down to the consequences of the transfer of electrons from one species to another. You have already caught a glimpse of that process as we stood together perilously deep inside the blast furnace in Reaction 4 and saw that O^{2-} ions transfer electrons to Fe^{3+} ions to bring about the reduction of the ore to the metal.

Tighten your intellectual seat belt. I intend to develop the very sparse view that oxidation is the loss of electrons and reduction is their gain. That is the austere message to take from this chapter, but I will cloak it in velvet. In each case, atoms might also get dragged around in pursuit of the electron as it migrates between species. However, in every case the fundamental act is the transfer of electrons from what is being oxidized to what is being reduced. I can express that even more bonily: *oxidation is electron loss, reduction is electron gain*. Hold on to that message throughout this section.

To see the truth of this 'electrons in transit' view we need to stand back a little. You might already have noticed the first clue: oxidation never occurs without reduction, and vice versa. Oxidation and reduction are the Tweedledum and Tweedledee, the Castor and Pollux, of chemistry. There is never one without the other. For instance, when iron ore is reduced by carbon monoxide in a blast furnace, the carbon monoxide is oxidized to carbon dioxide. Because oxidation is always accompanied by reduction chemists take the view that it is inappropriate to speak of an oxidation reaction alone or of a reduction reaction alone. Instead, they think of them jointly as a reduction–oxidation reaction and speak of them as a 'redox reaction'. An oxidation reaction is like one hand clapping; so is a reduction reaction; jointly two hands clap as a redox reaction.

This point, that two hands need to clap, should be easy to understand once you accept that redox reactions are electron transfer reactions, for there cannot be electron donation without electron acceptance. You saw exactly the same reciprocity, the same need for two hands to clap, in the discussion of neutralization and proton transfer in Reaction 2: an acid, a proton donor, cannot donate into a vacuum: it needs a base, a proton acceptor, to participate in the transfer.

Burning metal

Let's look at the oxidation and reduction reactions we have already encountered, but do so in the expectation of winkling out the fact that they all take place by the transfer of electrons. A good place to start is the combustion of magnesium (Reaction 3).

When magnesium burns it produces magnesium ions that end up as magnesium oxide, MgO: that is, as the metal burns, Mg atoms become Mg^{2+} ions, **1**. Clearly, this step involves electron loss, so according to the new definition in terms of electron loss it is an oxidation. As the metal burns, the oxygen of the air is converted to oxide ions, O^{2-}, **2**. This step must involve the transfer of elec-

trons to the molecule to create the negative charge of the ions (as well as break the O_2 molecule into atoms). So, the reaction of an O_2 molecule must involve the transfer of electrons to it: according to the new definition, oxygen has been reduced.

You should now also be able to appreciate that 'oxidation', despite its name, need not involve oxygen: any electron acceptor can play the role of oxygen. I remarked in Reaction 3 that when magnesium burns in air it also combines with the atmospheric nitrogen, which is one reason why the reaction is so vigorous (normally nitrogen dilutes the oxygen; here it is a fully collaborating partner). The product is magnesium nitride, which is built from Mg^{2+} ions and nitride ions, N^{3-}. Because Mg has lost electrons to become Mg^{2+}, according to the new definition the process is an oxidation of magnesium even though no oxygen is involved. Likewise, according to the new definition of reduction as electron gain, nitrogen as N_2 molecules has gained electrons in the formation of N^{3-} ions, so it has been reduced.

Let me summarize where we are. Whenever a species loses electrons, regardless of where they have gone, we say that it has been oxidized. Whenever a species gains electrons, regardless of where they have come from, we say that it has been reduced. The migration of electrons from one species to another is a reduction–oxidation, or redox reaction. As we shall see, such reactions are of huge importance in chemistry, life, and the universe.

Agents in waiting

Let's see how this interpretation applies to the other reactions I have mentioned. You can probably see that the reduction of iron oxide fits in well: the Fe^{3+} ions of haematite (recall Figure 4.1) accept electrons and are turned into Fe atoms: they are reduced. The oxide ions of the ore surrender their electrons (two of them from each O^{2-} ion) and so, according to the new definition, are reduced to O atoms. That these O atoms attach to CO molecules is interesting, and shortly I shall invite you to think about it further, but for the moment there is a reduction (of Fe^{3+} to Fe) accompanied by an oxidation (of O^{2-} to O).

You should now be able to appreciate more fully the description I gave of the blast furnace in Reaction 4, with our eyes now alert to the migration of electrons. Indeed, it is engagingly ironical that such a huge device as a blast furnace is needed to encourage the migration of

FIG. 5.1 *Iron ion being reduced*

one of the smallest of all fundamental particles, an electron. The electrons that convert the Fe^{3+} ions to Fe atoms were originally on the oxide ions, O^{2-}. We have to conclude that the actual reducing agent, the agent that converted the iron ions of the ore to iron metal, consists of the oxide ions that surround the iron ions. You can now appreciate that O^{2-} ions are pervasive enemies in waiting, a dormant fifth column. They might lie for millennia, even aeons, side by side with Fe^{3+} ions, but carbon monoxide and high temperature jointly awake their potency. By eloping with the O atom that carried the extra electrons, carbon monoxide lets loose electrons from the O^{2-} ions. They spring onto the Fe^{3+} ions and thereby form the iron the furnace is designed to produce (Figure 5.1).

One consequence of thinking like this and realising that CO is just playing a scavenging role for O atoms is that it might be possible to use different scavengers. The current process has a heavy carbon footprint, because even under perfectly efficient conditions and even ignoring the CO_2 produced when the slag-removing limestone is heated and decomposes, 1.2 tonnes of CO_2 are produced for every tonne of iron. And the conditions are far from perfect. If CO could be replaced by another scavenging agent, and it proved cheap enough, then the carbon footprint of the industry could be reduced dramatically. Maybe you can think of one, and make an environmentally benign fortune.

Subtle shifts

There is another interesting wrinkle in this discussion. The conversion of CO to CO_2 certainly looks like an oxidation in the old-fashioned sense of reaction with oxygen, but what is happening in terms of electrons? In what sense, in the new language of electron transfer, is the attachment of O to CO an oxidation?

To answer this question, I need to help you to look closely at the formation of the bond between the C and O atoms. This is an important point, because the electron transfer characteristic of redox reactions is often a bit difficult to identify, especially if atoms have changed partners too and have thereby muddied the waters. I need to show you how to detect it, or at least show you how chemists identify it as they ponder their equations.

The point to appreciate is that for a bond to form when an O atom attaches to the C atom of CO in the process of forming CO_2, the incoming O atom needs to share electrons provided by the C atom. That is, when it attaches, the O atom acquires a share of electrons at the C atom's expense. When sharing takes place, electrons are transferred, partially at least, from the C atom of CO to the O atom, 3. In other words, the O atom is reduced and the C atom is oxidized. The attachment of O to CO is therefore indeed another redox process!

This discussion will also help to explain why the combustion of methane that I described in Reaction 3 is also a redox reaction in the sense of involving electron transfer. It is obviously an oxidation reaction in the old-fashioned sense of being a reaction with oxygen, but I need to show that it conforms to the modern definition in terms of electron transfer.

The explanation hinges on the fact that judging whether electrons have been transferred boils down to judging shifts in the sharing of electrons between atoms. Getting a greater share of the available electrons counts as electron gain, and therefore reduction. Losing a share, even partially, counts as electron loss, and therefore oxidation. There are always two hands clapping, but in some cases the clapping is only half-hearted.

Let's apply the last point to the combustion of methane, which I treated in Reaction 3. When methane burns, the four H atoms attached to the C atom in CH_4 are replaced by the two O atoms in CO_2. So, our problem is to understand why the replacement of H atoms by O atoms is in some sense an electron transfer process.

In a chemist's mind's eye, the argument runs as follows. In CH_4, the C atom has a greater share than the H atom of the electron pair used to form each C–H bond, so the C atom is slightly electron-rich at the expense of its partner H. The opposite is true of carbon dioxide. In CO_2, largely on account of the greater nuclear charge of an O atom compared to that of an H atom, the O atoms have a greater share than the C atom of the electron pairs used to form the C–O bonds. That is, the C atom is slightly electron-poor and the O atoms are electron-rich. You should now be able to appreciate that, because in the combustion reaction electron-rich C becomes electron-poor C, when CH_4 becomes CO_2 (Figure 5.2), the carbon in CH_4 has been oxidized (has lost electrons).

What has been reduced? What other hand is clapping? You should suspect that, because there is no other candidate, it is the oxygen. But to make sure that the generalization of the definition is proving appropriate you should check by examining shifts in the patterns of electron sharing. The O atoms in the original O_2 molecules share the bonding electrons equally. In CO_2 they have gained partial possession of carbon's electrons. Because the O atoms are now slightly richer in their share of electrons, they have indeed been reduced.

Don't forget the other product of the combustion reaction, water. A camel doesn't: it uses this oxidation product as an internal source of refreshment. In H_2O, electrons have been sucked slightly off the H atoms towards the electron-hungry O atom and the O atom is consequently slightly electron-rich. So the O atoms of the original O_2 molecules have been reduced. The H atoms originally in CH_4 partially lose their share of bonding electrons when C gets replaced by O, and have therefore been oxidized.

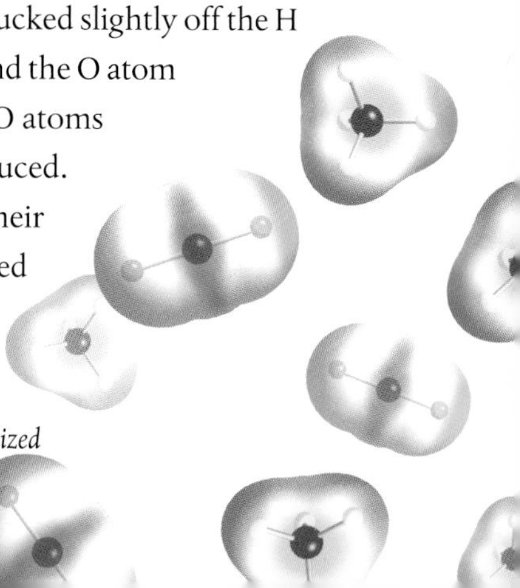

FIG. 5.2 *Carbon being oxidized*

The overall outcome of the combustion reaction is the partial loss of electrons from both carbon and hydrogen (their oxidation) in CH_4 and their partial gain by oxygen (its reduction) in O_2. Thus, the combustion of methane, and by extension, of any hydrocarbon, is a redox reaction in which the hydrocarbon is oxidized and oxygen is reduced.

Quiet revolutions

As I have stressed, neither oxidation nor reduction needs a substance of any particular identity to be present. All that needs to take place is for electrons to migrate from one species to another. If that happens, then whatever the substances involved, the reaction is classified as redox. All the reactions so far have involved flames of various kinds, but that is far from essential. Most redox reactions are quietly flame free; they are quiet revolutions.

A simple but important example is the reaction that occurs if you drop a piece of zinc into a solution of copper sulfate. You will find that the zinc becomes coated in a layer of copper and that some of the zinc dissolves (as Zn^{2+} ions). You can conclude that the zinc atoms, Zn, in the lump, having lost electrons, have been oxidized to Zn^{2+} ions. Moreover, because the Cu^{2+} ions in the solution have been converted to copper atoms, Cu, by gaining electrons, the ions have been reduced (Figure 5.3). This redox reaction might seem rather dull and a waste of zinc, but you will see in Reaction 7 that it is the foundation of the emergence of modern communication systems.

Everyday life

This discussion underlies not only the unification of many seemingly disparate types of reactions into a single family but also the importance of redox reactions for everyday and commercial life.

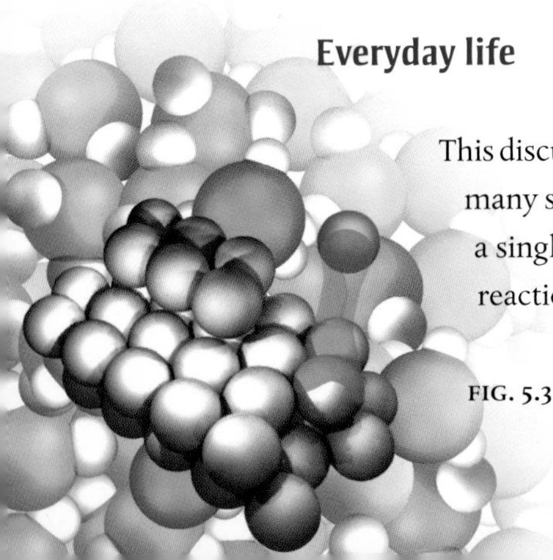

FIG. 5.3 *Zinc reducing copper ions*

You have seen that heavy chemical industry depends on redox reactions for producing iron (Reaction 4). Electrolysis reactions (Reaction 6) are redox reactions; for instance, in the production of aluminium, electrons are transferred to the Al^{3+} ions of bauxite to convert it to Al atoms. Combustion reactions (Reaction 3) are redox reactions, so our transport and much of our energy production in power stations are impelled by them. When you make a journey, your progress is due to the operation of redox reactions as the fuel burns, with electrons transferring from one species to another and in the process turning the wheels of your vehicle. The reactions in batteries that drive all our portable electronic devices are redox reactions (Reaction 7); corrosion is the outcome of redox reactions (Reaction 8).

Your metabolism is driven by redox reactions, because the consumption of foods is their oxidation, so at heart eating is mastication followed by incorporation by redox reaction. (You will see in Reaction 27 that proton transfer is an ally of electron transfer when it comes to digestion, with protons helping electrons on their way.) Foods are basically fats, proteins, and carbohydrates in proportions that depend on your choice from the menu. Fats are largely hydrocarbons and are the digestible analogue of gasoline and diesel fuel, 4. Like fuel in vehicles, they are readily stored, with the gut or hump taking the place of the tank. Proteins are rather special and too precious just to burn, and I deal with them in Reaction 27. You can think of carbohydrates as hydrocarbons that have already been partially oxidized by the addition of O atoms to give molecules such as glucose, $C_6H_{12}O_6$, 5. Their partial

oxidation and consequent possession of O atoms in their structures means that they are readily soluble in water; thus they are mobile and hence readily accessible sources of immediate energy. In each case, be it diesel fuel, burger, or chateaubriand, the reaction that takes the input to the output of carbon dioxide and water is a redox reaction. Thus, your daily life, your deeds, and all your thoughts are driven by the migration of electrons from what you ingest to the oxygen that you inhale.

Electric Occurrence

ELECTROLYSIS

lectrolysis makes use of electric currents, a stream of electrons, to bring about chemical change. It puts electricity to work by using it to break or form bonds by forcing electrons on to molecules or sucking electrons out of them. Electrolysis is an application of the redox processes I described in Reaction 5, where I showed that reduction is the gain of electrons and that oxidation is their loss. All that happens in electrolysis is the use of an external supply of electrons from a battery or other direct-current (DC) source to push them on to a species and so bring about its reduction, or the use of the electron-sucking power of a battery to remove them from a species to bring about its oxidation. Electrolysis, in other words, is electrically driven reduction and oxidation.

In fact, the process is rather broader than just forcing species to accept or give up electrons because, as I have hinted, molecules might respond to the change in their number of electrons by discarding or rearranging atoms. For instance, when water is electrolysed, the H–O bonds of the H_2O molecules are broken and hydrogen and oxygen gases are formed. When an electric current is passed through molten common salt (sodium chloride, NaCl), metallic sodium and gaseous chlorine, Cl_2, are formed. Electrolysis is a major technology in the chemical industry, for among other applications it is used to make chlorine, to purify copper, and to extract aluminium.

To bring about electrolysis, two metal or graphite rods, the 'electrodes', are inserted into the molten substance or solution and connected to a DC electrical supply. The electrons that form the electric current enter the substance through one electrode (the 'cathode') and leave it through the other electrode (the 'anode'). A molecule or ion close to the cathode is forced to collect one or more electrons from that electrode and be reduced. A molecule or ion close to the anode is forced to release them to that electrode and thereby become oxidized.

Copper bottomed

A reasonably simple first example is the purification of copper. Its purification is hugely important for electrical applications because its conductivity is severely impaired if impurities are present because, like boulders in a stream, they inhibit the electron flow. To visualize the process, think of one electrode, the anode, as impure copper metal and the other electrode, the cathode, as pure copper. Both electrodes are immersed in a solution of blue copper sulfate in water. Copper sulfate solution consists of positively charged copper ions, Cu^{2+} (Cu denotes copper, *cuprum*) and negatively charged sulfate ions, SO_4^{2-}, **I**, all of them free to migrate through the water. Sulfate ions are sturdy, and are just inert, disinterested spectators in all that follows.

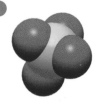

As usual, let's shrink to the size of a molecule, stand together close to the edge of the cathode, and watch what is going on there (Figure 6.1).

REACTIONS

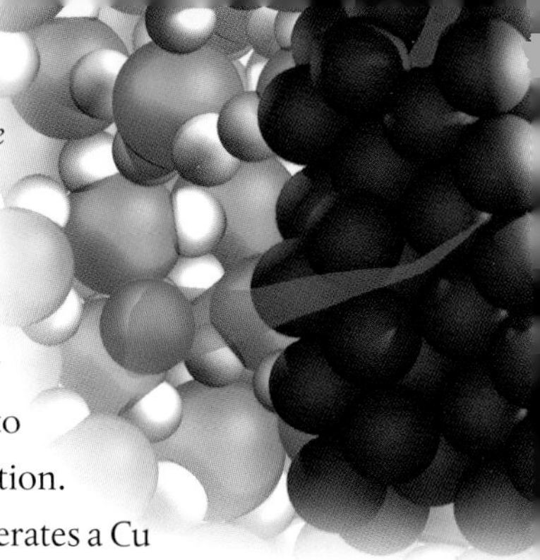

FIG. 6.1 *Action at the cathode*

The electrons are pushed on to the cathode by the external source and ripple through the metal. We see them seep from the pure copper of the electrode on to a Cu^{2+} ion that happens to have wandered inshore and be nearby in the solution. That transfer neutralizes the ion's charge and generates a Cu atom that attaches to the electrode. In the language of redox reactions (Reaction 5), the Cu^{2+} ions are reduced to Cu atoms by the incoming stream of electrons. As electrolysis continues, we see that more and more copper atoms are generated in this way all over the electrode, which grows as pure copper is deposited.

We now wade together through the blue solution across to the anode and inspect what is happening there (Figure 6.2). The anode is where electrons are being sucked out by the external battery. We see electrons are being drawn off the copper atoms that make up the impure copper metal electrode and scurry away along the external circuit. That is, in less picturesque redox language, Cu atoms are being oxidized to Cu^{2+} ions to satisfy the hunger of the external battery for electrons. The newly formed Cu^{2+} ions, no longer gripped to their copper neighbours by electrons, tumble into solution, making up the loss of Cu^{2+} ions that is occurring over at the cathode. As the process continues, we see the anode crumbling as its atoms are lost and the entrapped impurities are released and deposited below it as a sludge. Actually, the sludge might be valuable as it contains precious metals, enough perhaps to offset some of the cost of the electricity, but that is another story.

I wrote 'tumble into solution'. That is not quite right: it is more like hijacking by water. If we watch carefully, we see that as electrons are removed from a Cu atom on the surface of the electrode, water molecules cluster around it. The partial negative charge on the O atom (remember my *Preliminary remark* on water)

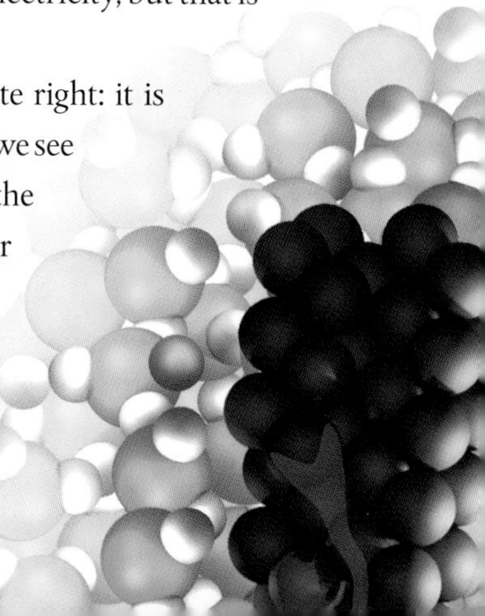

ELECTRIC OCCURRENCE

49

FIG. 6.2 *Action at the anode*

of nearby water molecules detects the emerging positive charge of the ion; the molecules respond to it, surround it, and carry it off into solution as a hydrated (water-surrounded) ion.

Breaking water

Now for the electrolysis of water itself. This process is a bit more complicated because the transfer of electrons results in the rupture of oxygen–hydrogen bonds, so more is going on than when an atom or ion just picks up or loses a charge. I shall suppose that the electrodes are made from platinum, a noble (chemically aloof) metal that does not undergo chemical change in the course of the electrolysis but, as you will see, unlike some nobles is not entirely idle. A further point is that water itself is a very poor conductor of electricity and very little current flows through it unless ions are present to carry charge from one electrode to the other. Let's forget these complications and think of water in the manner described in Reaction 2, as consisting mostly of H_2O molecules with a smattering of H_3O^+ and OH^- ions (as in Figure 2.3). These ions are always present as a result of the incessant proton transfer processes taking place between the molecules.

The electrons in the electric current from the external DC source stream onto the cathode. As we watch, we see it becoming bloated with electrons and therefore negatively charged (Figure 6.3). We also see a shoal of positively charged H_3O^+ ions swarming through the water towards the attracting charge of the electrode. As soon as one H_3O^+ ion comes into contact with the metal electrode we see an electron jump across, envelope the ion, and turn it into an unstable H_3O uncharged molecule.

This is where the platinum plays a role. Just as soon as it is formed, we see the H_3O molecule discard a single hydrogen atom, so becoming a much more stable H_2O molecule. That mol-

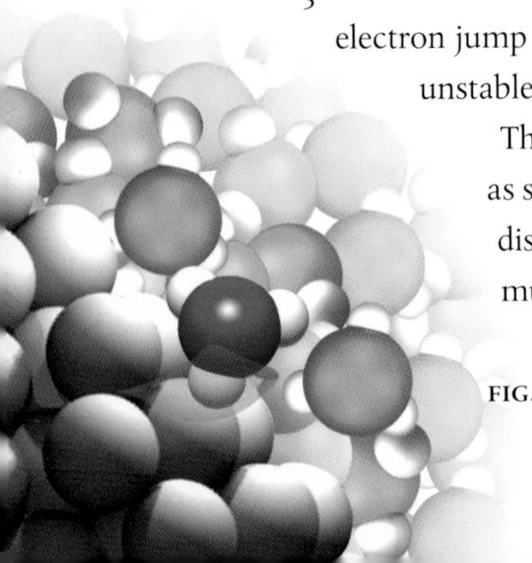

FIG. 6.3 *Reduction at the cathode*

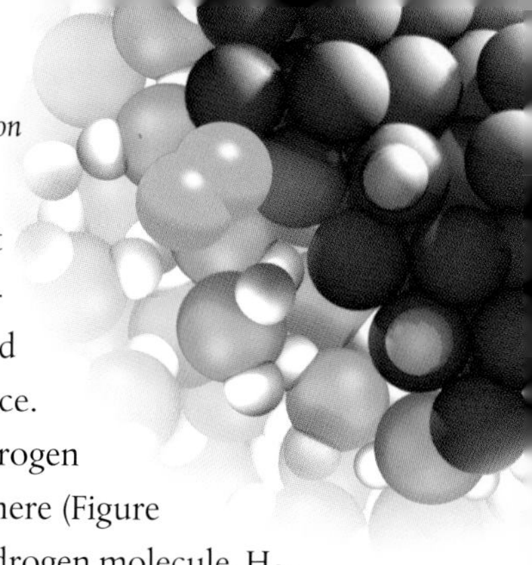

FIG. 6.4 *Hydrogen formation*

ecule joins the myriad others jostling around it
and gets lost in the crowd. The discarded hydrogen atom attaches loosely to a platinum atom and
is relatively free to skitter over the metal's surface.
As it does so, we see it bump into another hydrogen
atom that was formed in a similar event elsewhere (Figure
6.4). The two atoms stick together to form a hydrogen molecule, H_2,
which is released from the surface. Although we are observing only
one, vast numbers of events like it are happening all over the surface
of the electrode, and the molecules so formed congregate into a bubble of hydrogen gas, which rises to the surface.

We now splash through the water across to the other electrode,
the anode. There, we see electrons being sucked out of the electrode
by the external source, which needs them for it to continue to supply
current to the cathode (Figure 6.5). As a result of the loss of electrons,
the anode becomes slightly positively charged. We see a shoal of OH^-
ions in the water swarm across to the electrode's surface, attracted by
its positive charge. Then as we go on watching, a complicated dance
takes place as OH^- ions dump their extra electron into the platinum
electrode. The electrons ripple through the metal on their way to the
external circuit and are lost to sight. Some of the O and H atoms separate and migrate over the surface. As they hop across it, they collide
with other atoms produced in similar events elsewhere. We see two
O atoms combine to form an O_2 molecule. Two other O atoms mop
up the four H atoms in a series of steps to form two H_2O molecules
that just wander away and become part of the surrounding water.
Somewhere else we catch sight of an H atom colliding with
an intact OH group to form an H_2O molecule that also
joins the throng. Myriad events like those are taking
place all over the surface, and the O_2 molecules
congregate and bubble off as oxygen gas.

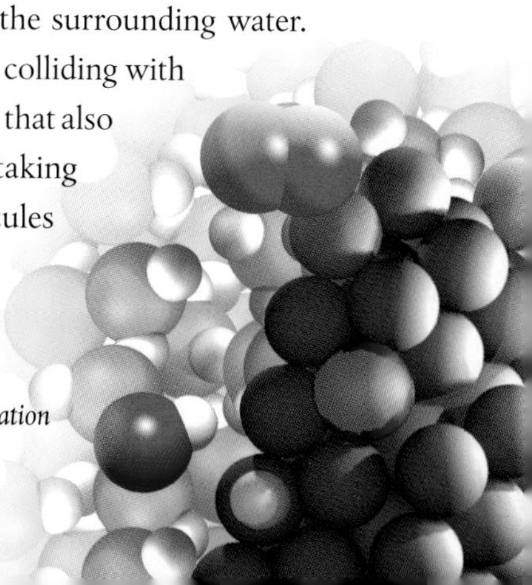

FIG. 6.5 *Oxygen formation*

Undermining Napoleon

Finally, let's consider the extraction of aluminium, Al. Aluminium was a very rare and expensive element when it was first identified but is now hugely important in construction and technology. It was so rare originally that it is said that Napoleon used aluminium plates for his favoured guests, the rest having to make do with gold.

Then along came the electrolytic process for extracting aluminium from its principal ore, bauxite, an oxide of aluminium of formula Al_2O_3, and changed the world. This procedure was developed by the young Charles Hall (1863–1914) in the United States and Paul Hérault (also 1863–1914) in France at about the same time (1888), and almost overnight turned aluminium from a curiosity into a necessity.

The electrolysis of bauxite is a bit more complicated than the processes I have already described, because the oxygen it contains must be eliminated and not allowed to react with the newly formed aluminium metal. For this reason, carbon (graphite) electrodes are used, so that the oxygen released in the electrolysis is converted into carbon dioxide, CO_2, and allowed to escape into the atmosphere out of one harm's way but into an ecological other. The carbon is acting in much the same role as in a blast furnace (Reaction 4), where its main job is to carry off the oxygen of the iron ore and take it out of reach of the iron. Apart from that consideration, the essence of the Hall–Hérault process is to dissolve bauxite into molten cryolite, a fluoride of aluminium, AlF_3. Then electrons are pushed from a cathode on to the Al^{3+} ions present in the melt to reduce them to Al atoms (Figure 6.6).

Because three electrons have to be supplied to convert each Al^{3+} ion into an Al atom, the process is a great consumer of electricity. Moreover, because three CO_2 molecules are generated in the electrolysis for every four Al atoms produced, corresponding to the release of 1.2 tonnes of carbon dioxide for 1

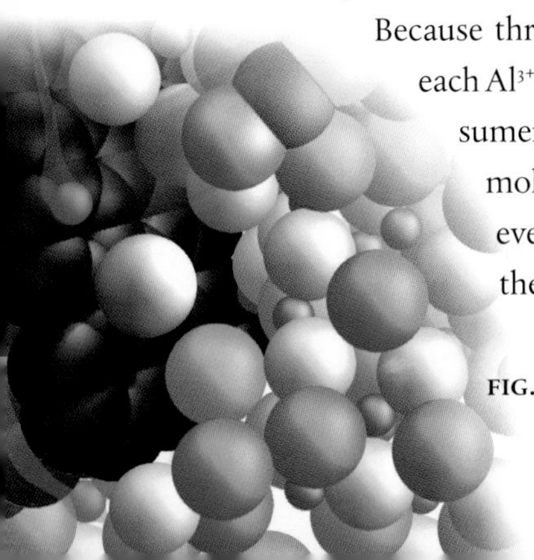

FIG.6.6 *Aluminium formation*

tonne of aluminium produced, the process has a heavy carbon footprint. There are therefore important economic considerations about where to locate aluminium sites, which ideally should be close both to sources of bauxite (Jamaica, for instance) and to cheap, carbon-low, typically hydroelectric, sources of electricity. The two requirements are often in conflict. It can therefore also be appreciated that the recycling of aluminium, which does not require the injection of so many electrons because the neutral Al atoms are already present, is economically and environmentally highly attractive.

7

The Generation Game

ELECTROCHEMISTRY

You already know, if you have read Reaction 6, that an electric current is a stream of electrons. If you have also read the section on redox reactions (Reaction 5), which you should, in preparation for this account, then you will also know that in a redox reaction electrons are transferred from one species to another. Although it is now far too late, had you had that information 150 or so years ago, then you might have realized that if those species were at the opposite ends of a piece of wire, the transfer of electrons would then take place in the form of an electric current travelling along the wire and you would have invented the electric battery. All the batteries that are used to generate electricity and drive portable electrical and electronic equipment, from torches, drills, phones, music players, laptops, through to electric vehicles, are driven by this kind of chemically produced flow of electrons.

An early ancestor

One of the earliest devices for producing a steady electric current was the 'Daniell cell', which was invented in 1836 by John Daniell (1790–1845) of King's College, London in response to the demand in the nineteenth century of the then emerging technology of telecommunication for a steady, cheap source of electricity. I have already touched on the underlying reaction when I explained what happens when a piece of zinc, Zn, is dropped into a solution of copper sulfate (Reaction 5), and this section builds on that account.

In that reaction copper is deposited on the zinc and the copper sulfate solution gradually loses its colour as blue Cu^{2+} ions are replaced by colourless Zn^{2+} ions. As this reaction takes place, electrons hop from the zinc metal onto Cu^{2+} ions nearby in the solution. If we were to stand there watching, we would see electrons snapping across from the zinc to the Cu^{2+} ions wherever the latter came within striking distance of the zinc surface. There would be electron transfer, but no net current of electricity.

Daniell did what I outlined in the opening paragraph: he separated the zinc metal and copper ions, so that electrons released by zinc had to travel through an external wire to get to the Cu^{2+} ions. To achieve this separation, he immersed an earthenware pot containing a solution of zinc sulfate and a zinc rod in a solution of copper sulfate that also contained a copper rod. An unglazed earthenware pot is porous, so water can penetrate into it and ions dissolved in the water can migrate though it and ensure that the reaction doesn't grind to a halt as electric charges build up.

As usual, let's imagine shrinking to the size of a molecule and diving into the solution in the inner pot and making our way to the edge of the zinc electrode (Figure 7.1). We see Zn atoms letting slip electrons into the rest of the metal

FIG. 7.1 *Zinc oxidation at the zinc electrode*

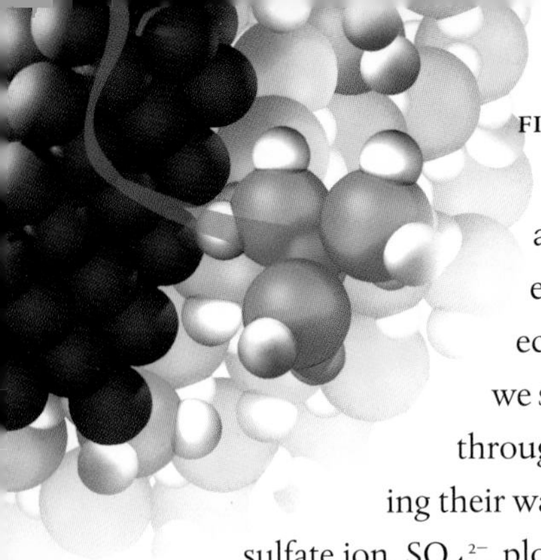

FIG. 7.2 *Copper ion reduction at the copper electrode*

and wandering off into solution as Zn^{2+} ions, each one surrounded by a huddle of H_2O molecules. If we track across to the edge of the pot, we see Zn^{2+} ions released in earlier events wriggling through the convoluted pores of the pot and making their way into the outer vessel. Occasionally even a big sulfate ion, $SO_4{}^{2-}$, plops out into the solution on our side of the barrier, but these are big ions and few manage to make the journey from the other side. If we manage to wriggle through the pores ourselves, we emerge in the blue solution of the outer pot, which is full of Cu^{2+} and $SO_4{}^{2-}$ ions. Meanwhile, the liberated electrons travel through the outer circuit, driving whatever happens to be connected in the circuit, and enter the copper electrode in the outer pot. We swim across to see what is going on there.

At that electrode we see electrons seeping out of the copper rod and, if there is a Cu^{2+} ion within reach, snapping onto it, reducing it to a Cu atom (Figure 7.2). We see that these atoms either stick to the electrode or just congregate together on the surface and fall off as sludge.

Overall, the same reaction has occurred as when zinc is simply dipped into copper sulfate solution, but because the sites of oxidation (Zn to Zn^{2+}) and reduction (Cu^{2+} to Cu) are now separated spatially, in this arrangement the reaction has generated an electric current.

Beyond pots

Modern portable electronic technology has moved on from earthenware pots (but their descendant ceramics are very much in vogue). Nevertheless, the principles remain the same. In the currently widely used lithium–ion battery, electrons are released from lithium atoms, Li, as they change into Li^+ ions, and those ions drift away from the electrode. Pure metallic lithium is not used for the electrode in view of a number of difficulties, including the risk of fire. Instead, a range

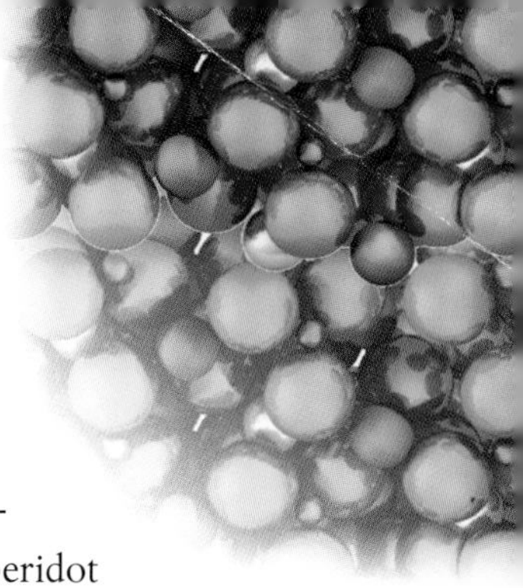

FIG. 7.3 *Olivine*

of compounds are used in which Li atoms nestle safely but accessibly inside a solid matrix.

If we manage to get inside one such battery we find that one electrode is based on the structure of the mineral olivine, which takes its name from the olive-green of the parent material, as in the gemstone form peridot (Figure 7.3). We can see from its structure that in olivine the Li atoms are imprisoned within a framework built from Mg^{2+}, Fe^{2+}, and silicate ions, SiO_4^{4-}, I. Like prisoners in real gaols, the imprisoned Li atoms are less hazardous than when present as raw lithium metal. As we watch, we see a Li atom release an electron into the external circuit and become a Li^+ ion.

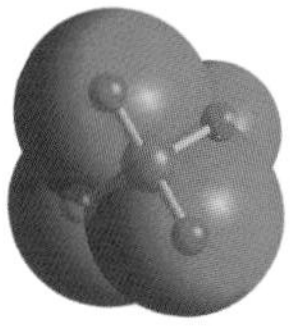

Getting across to the other electrode through the murky organic medium between the electrodes is more of a problem than getting through the water of the Daniell cell, but we finally make it and on arrival find that the electrode is principally graphite. Graphite is a form of pure carbon in which carbon atoms lie linked together in hexagons to form a planar chicken-wire-like network, with thousands of layers stacked on top of each other (Figure 7.4). The layers are stuck together only weakly and it is easy for Li^+ ions to wriggle between the layers and lie trapped there. The layers are quite good conductors of electricity, and as we watch, we feel a tremble in the sheets, then see a sudden local convulsion as an electron arrives from the external circuit and attaches to a Li^+ ion, so forming a Li atom.

When the cell is being charged, an electric current is forced to travel from the graphite electrode to the olivine electrode by connecting the battery to an external powerful source. At the graphite electrode a Li atom is forced to surrender an electron to the external circuit, is converted into a Li^+ ion, and is expelled into the inter-electrode medium. It migrates across to

FIG. 7.4 *Lithium in the graphite electrode*

the olivine electrode. In the neighbourhood of the olivine electrode, Li^+ ions are converted to Li atoms by electrons pushed on to them by the external source, and enter the solid structure.

When the cell is being discharged and running the equipment in its external circuit these processes are reversed. A Li atom in the olivine releases an electron into the external circuit. A Li^+ ion close to the graphite accepts an electron, is converted to a Li atom, and wriggles between the sheets of carbon atoms.

Numerous variations on this theme have been developed because the generation of electric currents from light weight systems is so important for the modern world, but all of them are based on the encouragement of electron transfer and its capture by separating the source of electrons from the sink into which they flow.

Cold furnaces

A variation on the theme is the 'fuel cell', a device first constructed by Sir William Grove (1811–1896) in 1840 on the basis of ideas proposed by the German chemist Christian Schönbein (1799 –1868) shortly before. A fuel cell works in much the same way as the cells I have already described, using redox reactions to generate a flow of electrons. However instead of the reagents being sealed in at manufacture they are supplied continuously from an external source, like fuel to a furnace, as the reaction proceeds.

In a primitive fuel cell, the only kind I shall consider, the fuel is hydrogen and the oxidizing agent is oxygen from the air. The hydrogen gas, H_2, flows over one electrode, which is typically platinum. As we look at the surface of the electrode, which looks like a cobbled street of platinum atoms, Pt, we see an incoming H_2 molecule attach to it and break apart into H atoms (Figure 7.5). These atoms are only loosely attached to the surface, and we see

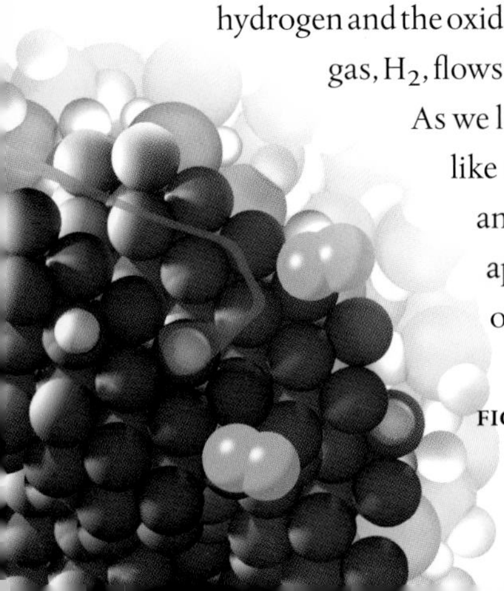

FIG. 7.5 *Hydrogen oxidation at platinum*

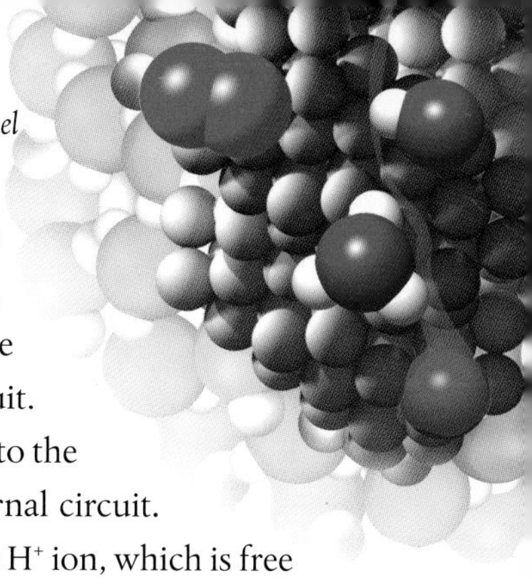

FIG. 7.6 *Oxygen reduction at nickel*

them skidding about over it. If we watch one carefully, we see that it might linger at a Pt atom, and find that its electron has slipped off it into the platinum metal on its way to the external circuit. Each atom is able to release its single electron into the metal and send it on its way through the external circuit. The loss of an electron turns the H atom into an H^+ ion, which is free to leave the electrode and migrate through the intervening medium to meet its fate at the other electrode.

The other electrode, which is typically nickel, Ni, is the site of reduction of oxygen. We move across to it to see what is happening there (Figure 7.6). We see the O_2 molecules attaching to the nickel surface, breaking apart, and forming O atoms. Then a proton, riding on an H_2O molecule as H_3O^+, gallops into the surface and accepts an electron from it. That proton becomes an H atom, slips off its carrier, and attaches to the surface. Once there, loosely attached, it skitters across the surface, encounters an O atom, and becomes OH. Soon another H atom formed similarly elsewhere on the surface, skitters across it and bumps into the OH anchored to the surface. We see an oxygen–hydrogen bond form and H_2O produced. That molecule wanders off, in due course to be drained from the cell. If the cell is on a spacecraft, then like the fatty hump on a camel (Reaction 3) the product is a camel-like contribution to the water supply on board.

Overall, hydrogen has combined with oxygen to give water, but in the fuel cell arrangement, the process has been harnessed to drive electrons through a circuit. Instead of combustion and the wild release of energy as heat (Reaction 3), there is the controlled, almost cold release of energy as electricity.

All manner of different configurations of fuel cells have been and continue to be developed, some using hydrocarbon fuels, such as the methane of natural gas, in place of hydrogen, and having electrode materials and inter-electrode media of increasing sophistication. The underlying mechanism is the same in every case, and the same as in an ordinary battery. Oxidation takes place at one electrode and reduction takes place at a spatially separated electrode in an overall redox reaction. The electrons released in the former have to travel through an external circuit, where their passage is harnessed to do work, before they arrive at the other electrode to complete the redox reaction that drives the entire arrangement.

The Death of Metal

CORROSION

As in life, so in redox reactions: some are good and some are bad. Corrosion is one of the evil among redox reactions. It is the unwanted oxidation of a metal that cuts short the lifetimes of steel products such as bridges and vehicles. Replacing corroded metal parts costs industry and society a huge amount each year. Understanding it helps us to find ways to prevent it. Not all corrosion, however, is unwanted: the green patina of copper roofs is often sought and can be beautiful; the induced oxidation of aluminium in the presence of dyes can also be intentional and can bring graceful colour to a building.

I shall focus on the corrosion of iron, Fe (from the Latin *ferrum*), its rusting, as it is so common a way of death for our everyday artefacts. Iron rusts when it is exposed to damp air, with both oxygen and water present. In the process the Fe atoms of the metal are oxidized—lose some electrons—and become Fe^{3+} ions. These ions pick up some oxide ions, O^{2-}, and are deposited as the red–brown oxide, Fe_2O_3.

> *Pedant's point.* More precisely, the oxide is 'hydrated', in the sense that H_2O molecules are incorporated into the structure, with about one H_2O molecule for each Fe_2O_3 unit.

The corrosion of iron is very much like its reversion to the ore, which is also typically Fe_2O_3, from which, with so great an effort and all the expensive and energy-intensive, environmentally invasive fury of a blast furnace, it was originally obtained (Reaction 4). In the process of forming Fe^{3+}, the oxygen of the air, the oxidizing agent, is converted to water. The hydrogen atoms needed for the formation of H_2O molecules from O_2 molecules are scavenged from the surrounding solution, especially if it is acidic and rich in hydrogen ions.

Life on the edge

I shall now show you the reaction in more detail and try to lead you into appreciating visually what is going on inside a small droplet of water on the surface of a sheet of rusting iron. Although rusting is rarely thought beautiful, there is a beauty and subtlety in the choreography of the atomic events that underlie its formation. As usual, you should imagine shrinking to the size of a molecule, plunging below the droplet's surface, and descending diver-like through the densely agitating, bustling, tumbling water molecules. You descend until you stand beside me on the hard surface of the virgin metal, among the rocky outcrops of iron atoms and the swirling eddies of water molecules.

From where we made metal-fall we strike out together towards the edge of the droplet, pushing through the water molecules (Figure 8.1).

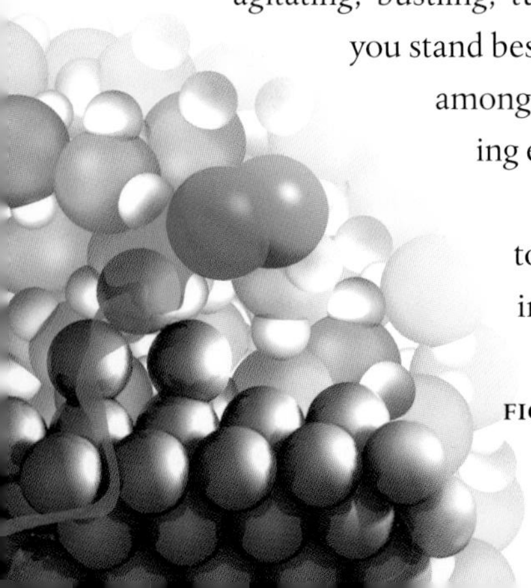

FIG. 8.1 *Reaction at the edge*

Near the edge we see oxygen molecules, O_2, splashing in from the air beyond and wriggling into the depths between the water molecules. Then we see violent electronic action. When a shark-like O_2 molecule reaches the iron surface it bites off two electrons from an Fe atom on an unwary outcrop. With two of its electrons gone, Fe has become Fe^{2+}. It begins to fall away from the outcrop and starts to be surrounded by a clustering shell of H_2O molecules.

The newly formed Fe^{2+} ion doesn't live for long, for the atoms on the edge of the droplet have a charmed life. As soon as the ion is formed, even as it is in the process of forming, we feel a ripple of electric current beneath our feet as electrons surge through the metal. They come from Fe atoms lying below the less oxygen-rich region of the centre of the droplet. Those atoms give up electrons and become Fe^{2+} ions. The electrons they have released migrate to the edge, stick to the Fe^{2+} ions there, and convert them back to Fe atoms again. So, it is the Fe atoms in that central region, behind our backs in the oxygen-poor region at the centre of the droplet, that are converted to Fe^{2+}. That is where the corrosion bites.

Let's go back a moment to when a shark-like O_2 molecule snipped electrons off Fe atoms at the edge of the droplet. Those electrons convert the invading O_2 molecules into two O^{2-} ions. But as soon as a virulent little O^{2-} ion is formed, it immediately snatches protons from H_3O^+ ions that happen to be in its vicinity. It and they become H_2O molecules, and wander away, lost and anonymous in the crowd of water molecules already present.

Death in the deep

You have seen the action at the edge of the droplet, so we now splash back through the water molecules to see what is happening towards the inner part of the droplet where the metal is being eaten away. Remember that the Fe atoms there had given up electrons to the atoms

FIG. **8.2** *Reaction at the centre*

at the edge, and Fe atoms on outcrops near the centre of the droplet have become Fe^{2+} ions. As these ions form they detach from the surface and become surrounded by H_2O molecule. These ions have adventures to come, for there are secondary processes that lead to the formation of rust itself.

We get back to the centre in time to see the newly formed Fe^{2+} ions drifting away into the surrounding water, leaving tiny pits in the surface of the iron (Figure 8.2). However, like real swimmers in real shark-infested waters, these Fe^{2+} ions do not live for long. There are cruising O_2 sharks even in this oxygen-poor region of the droplet. These O_2 molecules are, as usual, electron hungry, and are able to snip electrons off even these ions, and convert an Fe^{2+} ion into Fe^{3+}.

As we watch, we see each Fe^{2+} ion become an Fe^{3+} ion and the electron-enriched O_2 molecule fall apart into two O^{2-} ions. There is a big difference here, though, compared to what happened over at the edge of the drop. Here, very highly charged ions, Fe^{3+}, form right next to O^{2-} ions, and we see them clump together, drawn together by the strong attraction of opposite charges (Figure 8.3). As the ions converge to form a lump of oxide, they entrap H_2O molecules and various impurities in the growing solid mass. These particles of dirty, hydrated iron oxide fall around us like russet rain onto the already pitted surface as rust.

We have to fight our way to the surface of the droplet as best as we can, lest we become just one of billions of impurities buried in the vaccumulating rust.

Combating corrosion

Back at the surface, we can report what we have learned and seek to combat corrosion by using our understanding of what is going on in the

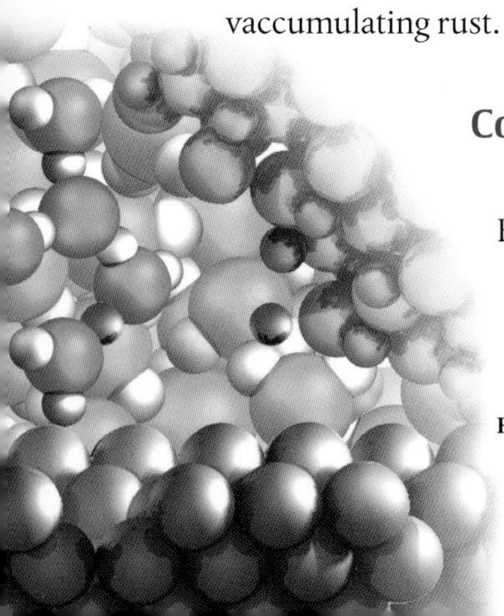

FIG. **8.3** *Rust formation*

FIG. 8.4 *Galvanized protection*

depths. The simplest way to prevent corrosion is to protect the surface of the metal from exposure to oxygen by cladding it in armour-plating. A simple form of armour plate is paint. Painting seals the surface away, and corrosion is prevented. One hazard remains: the paint might get scratched, its molecules and pigment scraped away, and the metal below revealed and opened to attack. A more sophisticated approach that overcomes this problem is to armour-plate the iron with zinc, Zn. This is the process of 'galvanizing' the metal.

Let's descend together into the depths again to watch what happens when a scratch scrapes away the layer of Zn atoms of a galvanized iron surface and exposes the Fe atoms beneath (Figure 8.4). As before, we feel the trembling of the surface as electrons ripple through it and travel across to the edge of the droplet. But look! Instead of an Fe atom becoming an Fe^{2+} ion, we see a nearby Zn atom release two electrons and convert the incipient Fe^{2+} ion back to an Fe atom before it has time to drift away and leave a pit. Thus, the zinc, not the iron, is oxidized and the metal artefact survives.

When it is not possible to galvanize large metal structures, such as ships, underground pipelines, big storage tanks, and bridges, 'cathodic protection' can be used instead. For example, a block of metal more willing to release electrons than iron, typically zinc or magnesium, can be buried in moist soil and connected by a conducting cable to an underground pipeline. The block supplies electrons to the iron as it releases its electrons to oxygen. These incoming electrons turn the incipiently forming Fe^{2+} ions back into Fe atoms, so preserving the integrity of the iron structure but at the expense of the oxidation of the block. The block, which appropriately is called a 'sacrificial anode', protects the pipeline and is relatively inexpensive to replace. For similar reasons, vehicles generally have 'negative ground' systems as part of their electrical circuitry. That is, the body of the car is connected to the anode of the battery, which is constructed from an electron-liberating metal. The decay of the anode in the battery is the sacrifice that helps preserve the vehicle itself.

Civil Partnerships

LEWIS ACID—BASE REACTIONS

One of the most remarkable chemists of the twentieth century, Gilbert Lewis (1875–1946), who died in a rather peculiar event involving cyanide (which will figure further in this account) took the story of acids and bases that I described in Reaction 2, extended its reach, and thereby captured a further huge swathe of chemical countryside. As I remarked in that section, chemists seek patterns of behaviour, partly because it systematizes their subject but also because it gives insight into the molecular events accompanying a reaction. Lewis contributed greatly to this enlargement of chemistry's vision, as I shall unfold in this section.

I explained in Reaction 2 how Lowry and Brønsted had extended Arrhenius's vision of acids and bases by proposing that all reactions between acids and bases involve the transfer of a proton (a hydrogen ion, H^+) from the acid, the proton donor, to the base, the proton acceptor. For instance, hydrochloric acid, HCl, can provide a proton that

sticks to an ammonia molecule, NH_3, 1, converting it into
NH_4^+, 2. The Lowry–Brønsted account of an acid–base
reaction involves a proton as an essential part of the defi-

nition: if protons aren't around, then Lowry and Brønsted are silent on whether a substance is an acid or a base.
There are, however, many reactions that resemble
acid–base reactions but in which no protons are transferred. I will give what might seem to be a rather
esoteric example, but it makes the point in a simple
and direct way, so please bear with me; you will soon
see the relevance of this presentation to everyday life.
The esoteric example I have in mind is a reaction in
which a boron trifluoride molecule, BF_3, 3, sticks to an ammonia
molecule to form NH_3BF_3, 4. This reaction clearly resembles the
proton transfer reaction in which H^+ attaches to NH_3 to form NH_4^+,
but with BF_3 playing the role of H^+.

Lewis brought these aspects together in a very
simple idea in 1923, at about the same time as Lowry
and Brønsted made their proposals. A base, he proposed, is any species that can use two of its electrons
to attach to an incoming species. That incoming species might
be H^+ or BF_3, for instance. A key feature of a base that Lewis had
in mind is its possession of a region where two electrons are concentrated but are not being used for bonding; they make up a kind
of electronic hot spot, 5 (the red, electron dense region). Chemists
call such a two-electron-rich region a 'lone pair'. The ammonia molecule has such a lone pair concentrated on the N atom, 5, and can
use the pair to form a bond to an appropriately receptive incoming
species, such as H^+ or BF_3. Lewis's proposal, therefore, is that a base is a species that has a lone pair of
electrons that can be shared with another species to form a bond between them. Notice that this

REACTIONS

definition doesn't use the word 'proton': any species with a suitably available lone pair can be a base.

Lewis's definition of an acid is also straightforward. Clearly, it has to be an entity that can attach to a base by sharing the latter's lone pair. Lewis proposed that that ability alone is sufficient for a species to be regarded as an acid. That is, according to Lewis, an acid is any species that can stick to a lone pair of electrons. Notice that the proton is not mentioned: any suitably accepting species can be called an acid; the proton is only one of a myriad possibilities.

6

At once we see that BF_3, which has only a wispy electron cloud on its B atom, **6**, and can accommodate an incoming lone pair, can be regarded as an acid, and that an acid–base reaction is nothing other than the sharing of an electron pair provided by the base. To function as an acid in this new sense, the species needs to have an empty patch in its electron cloud where the positive charge of the nucleus shines through and which can accommodate the lone pair of the incoming base. The proton is surrounded by emptiness (that is, it has no electron cloud), can stick to a lone pair supplied by any donor, and so functions as an acid.

Let me summarize. According to Lewis, a base (I shall now call it a *Lewis base*) is a species that can donate a lone pair to the formation of a bond. An acid (I shall now call it a *Lewis acid*) is a species that can accept a lone pair and so form a bond.

Simple complexes

Lewis's vision lets us see the proton in a different light. The proton itself is the species destined to share the lone pair provided by the base, so it, the proton itself, is the acid. This is in marked contrast to the approach I explained in Reaction 2, where the acid, such as HCl, is the source of the proton. According to the Lewis approach, the proton, like any other electron-pair acceptor, is the acid.

I need to take this discussion a little further before demonstrating the richness and considerable range of Lewis's approach. When a Lewis acid attaches to a Lewis base the outcome is the unit 'acid–base', with the acid attached to the base by a bond consisting of a shared electron pair provided by the base. This combination is called a 'complex'. So, NH_4^+, which we could write out as H_3N-H^+, is a complex of the Lewis acid H^+ and the Lewis base NH_3. Likewise, NH_3BF_3, which we could write out more revealingly as H_3N-BF_3, is a complex of the Lewis acid BF_3 and the Lewis base NH_3.

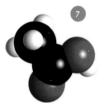

This terminology means that what in Reaction 2 I called an acid, such as hydrogen chloride, HCl, or acetic acid, CH_3COOH, 7, are, in this new terminology, complexes of the Lewis acid H^+ with the Lewis bases Cl^- and $CH_3CO_2^-$, 8, respectively. This interpretation gives new insight into the nature of a proton transfer reaction. You can now see that it is the transfer of a Lewis acid (H^+) out of one complex (such as CH_3COOH) to another base (such as NH_3), thus

forming a new complex (in this case, NH_4^+) and leaving behind a naked base ($CH_3CO_2^-$).

You already know (from Reaction 2) that a water molecule is just about all things to all molecules. It can act as a proton donor; it can act as a proton acceptor. I didn't mention the point in Reaction 5, but it can also act as an oxidizing agent and as a reducing agent. You shouldn't be surprised, therefore, that it can act as both a Lewis acid and a Lewis base. An H_2O, molecule, 9 has a two-electron-rich region on either side of its O atom and can use either of these lone pairs to act as a Lewis base. In fact, it might be unsettling to learn that when you drink a glass of water, you are swallowing pure Lewis base! That's not all. The regions of the molecule occupied by H atoms are electron-poor. That means that the positive charge of the nucleus of each atom can shine through the elec-

REACTIONS

tron cloud and be attracted to a region of high electron density, and specifically a lone pair. In other words, an H_2O molecule can act as a Lewis acid. If you don't like the thought that you commonly drink a pure Lewis base, you might like even less the realization that a glass of water is also a pure Lewis acid!

Here is a final unsettling insight. A water molecule is both a Lewis acid and Lewis base. When acid and base molecules are together, they form a complex. The trillions of water molecules in a glass of water are close together (recall Figure 1 in my *Preliminary remark* on water), so the water itself is one giant complex. That the molecules are stuck together is why water is normally a liquid, rather than a gas. So, when you drink the water, you are drinking a giant complex that snakes down your gullet.

Lewis's wonderland

I can now bring out the full glory of Lewis's approach and show how his huge net captures an extraordinarily wide range of different types of reactions and reveals that, despite superficial appearances, they all have a common feature. Some of these reactions you have already met in different categories, such as in redox reactions (Reaction 6) or precipitation reactions (Reaction 1). That doesn't mean that those classifications are defunct: it means that I am providing you with an enlarged vision of what is going on when one substance is transformed into another. With this broader perspective, you can think about the reactions in different ways.

Take, for instance, the reaction in a blast furnace (Reaction 4), to which I promised to return. You should recall that in a part of the process of reducing the ore to iron, a carbon monoxide molecule, CO, plucks out an O atom from inside an oxide ion, O^{2-}, and is converted into carbon dioxide, CO_2

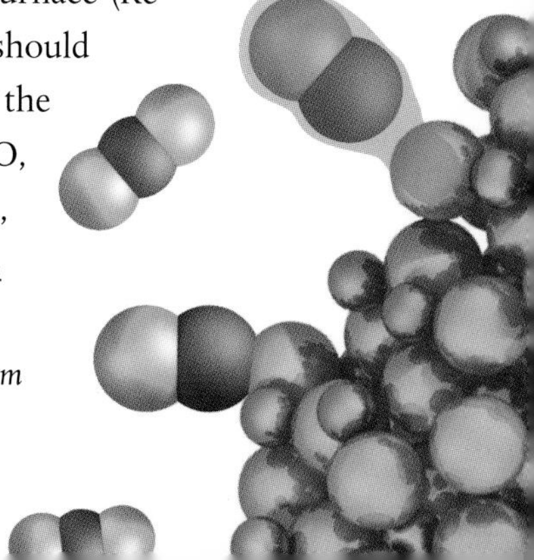

FIG. 9.1 *Carbon monoxide extracting an oxygen atom*

FIG. 9.2 *Carbonic acid formation*

(Figure 9.1). Not only is this an oxidation of CO, but it is also a Lewis acid–base reaction. If we look at the reaction through Lewis's eyes we see that an O atom has a gap for two electrons in its outer electron shell and is therefore a Lewis acid. We also see that CO has a lone pair of electrons on the C atom, which it can use to form a bond to the O atom. Thus, CO_2 is in fact a Lewis acid–base complex formed from the Lewis acid O and the Lewis base CO.

A molecule can change its spots. Thus, although CO_2 is an acid–base complex, it can behave as a Lewis acid if the opportunity arises. Let's watch what happens when CO_2 dissolves in water, to give what in one context is acid rain and another is soda water (Figure 9.2).

At first, a CO_2 molecule is simply surrounded by H_2O molecules, and we see it being jostled around like someone in a dense crowd. However, after a moment we see the lone pair of an H_2O molecule plunging into the C atom, form a bond to it, forcing the two oxygen atoms already there to angle back from the incomer. The outcome, after a bit of hopping around by a hydrogen atom, is a molecule

10

of carbonic acid, H_2CO_3, **10**. We have seen CO_2 acting as a Lewis acid, H_2O acting as a Lewis base, and carbonic acid, the outcome, being a Lewis acid–base complex.

11

Carbonic acid is the parent of the carbonates, ionic compounds based on the carbonate ion, CO_3^{2-}, **11**. You can think of the carbonate ion either as the Lewis acid–base complex carbonic acid that has acted as a proton donor and discarded two protons in a neutralization reaction or, more directly, as a complex of the Lewis acid CO_2 and the Lewis base, O^{2-}. Limestone is an impure form of calcium carbonate, $CaCO_3$, a stack of calcium ions, Ca^{2+}, and carbonate ions, CO_3^{2-}, and it is interesting to see what happens when limestone is heated. The

FIG. 9.3 *Calcium carbonate decomposing*

process is used commercially for the production of quicklime, which is calcium oxide, CaO, a compound used in the manufacture of cement and soda-lime glass. Limestone is also used as a scavenger of impurities in the blast furnace, as I mentioned in Reaction 4. The Ca^{2+} ions are just spectators of the sport that follows, and we need not consider them further.

Let's watch what happens when limestone is heated to around 800°C or so. We see the whole solid of Ca^{2+} and CO_3^{2-} ions vibrating increasingly rapidly (Figure 9.3). We also see that the CO_3^{2-} ions are also shaking internally. Soon the shaking is so violent that a CO_3^{2-} ion flings off a CO_2 molecule, which escapes through the crumbling edge of the solid, leaving an O^{2-} ion in its place. When that decomposition has occurred throughout the solid, we are left with quicklime. Note, though, that what we have witnessed is the decomposition of the Lewis acid–base complex, with the acid CO_2 being thrown off the base O^{2-} by the violence of thermal motion.

All the acid–base reactions that I described in Reaction 2 are also Lewis acid–base reactions. I have already remarked that in a proton transfer reaction, a Lewis acid (the proton, H^+) migrates from one complex (such as CH_3COOH) to a Lewis base (such as NH_3 or H_2O), and thereby forms a new complex (NH_4^+ or H_3O^+, respectively).

All the precipitation reactions that I described in Reaction 1 are also Lewis acid–base reactions, but this character is buried a little more deeply. To dig it out, let's consider the reaction in which silver chloride precipitates when solutions of silver nitrate and sodium chloride are mixed (just as I did in Reaction 1). You need to focus on the silver ions, Ag^+, and chloride ions, Cl^-, because the other ions present, the sodium and nitrate ions, are uninvolved spectators and remain unchanged.

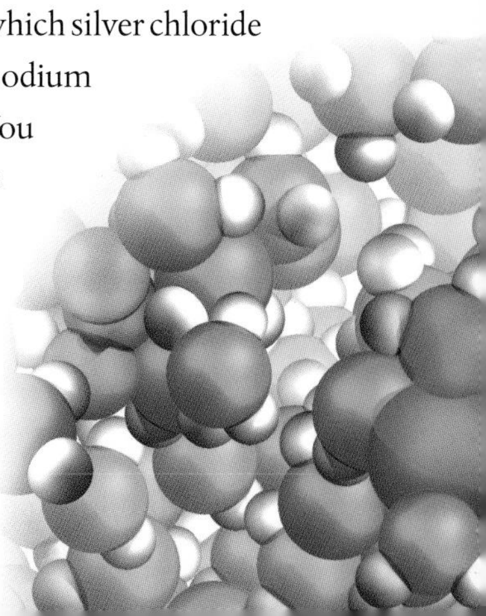

FIG. 9.4 *Silver chloride precipitating*

Let's watch the same events as in Reaction 1, but now through Lewis's eyes. Once again we see that the Ag^+ ions are hydrated (Figure 9.4). But Lewis whispers in our ear that a hydrated ion is actually an example of one of his complexes. The Ag^+ ion is capable of accommodating several H_2O molecules around it. Each H_2O molecule, which you know to be a Lewis base, uses a lone pair of electrons to attach to the central ion. That is, a hydrated Ag^+ ion is in fact the Lewis acid member of a complex, with the surrounding water molecules acting as Lewis bases.

When we catch sight of a hydrated Cl^- ion, we realize that something similar is going on. It is also part of a complex, but it is the base member. The ion uses its lone pairs to form weak bonds to the H atoms of the surrounding H_2O molecules: I pointed out earlier that an H_2O molecule can act as a Lewis acid.

When the solutions are mixed, the Cl^- ions shrug off their water molecules (that is, the complex decomposes), the Ag^+ do likewise, and the two liberated ions, the Lewis acid Ag^+ and the Lewis base Cl^-, form a new complex, the silver chloride, AgCl, that precipitates.

> *Pedant's point.* The new complex is not a single AgCl molecule, but an entire crystal (or at least a lot of little crystals), but the entire assembly can be considered to be a huge conglomerate of Lewis acid–base complexes.

One man's poison

Lewis acid–base reactions keep us alive. When we inhale air, the O_2 molecules, which are rich in lone pairs, form a complex with the Fe atom that lies in the four-fold heart of a haemoglobin molecule. The oxygen is thereby transported to other parts of the body, where it is released and used in the reactions that power us. Carbon monoxide,

FIG. 9.5 *Carbon monoxide and a fragment of haemoglobin*

CO, can also attach to the Fe atoms of haemoglobin (Figure 9.5; I have sketched in a representation showing the general layout of atoms
in a fragment of haemoglobin). If it does so, then it
blocks the entry of the O_2 molecule; thus Lewis acid–
base reactions can also lead to our suffocation.

And what of the reaction that killed Lewis? The cyanide ion, CN^-,
is a Lewis base because, like CO, it has a lone pair of electrons largely
centred on the C atom. You will see in Reaction 10 that many metal
atoms and ions can act as Lewis acids and form complexes with suit-
able Lewis bases. This is exactly what happened to Lewis, when the
CN^- ions that somehow, accidentally or intentionally, entered his
body attached to the Fe atoms in the molecules responsible for
channelling electrons from one species to another in the cascade
of reactions that power our bodies. The formation of the complex
inhibited the processes of electron transfer and terminated his
metabolism. Some people die by their own hand; Lewis died by his
own reaction.

10

Changing Partners

COMPLEX SUBSTITUTION

I shall now describe a special case of the Lewis acid–base reactions I introduced in Reaction 9. I showed there that a Lewis acid is a species that can accept a lone pair from another incoming species and form a bond to it, that a Lewis base is a species that provides that lone pair, and that the result of this sharing is a complex of the two species joined together by a chemical bond. The important special case I would like to share with you here is when the Lewis acid is a metal atom or ion, especially but not necessarily one drawn from the d-block of the periodic table (a 'transition metal'). The d-block consists of the elements that make up the skinny central rectangle of the periodic table. They include important constructional metals, such as iron, nickel, and copper, and also the chemically aloof 'noble' metals gold, platinum, and silver. The Lewis base that I focus on will be a molecule or ion that also has an independent existence in the wild, such as water, H_2O, or ammonia, NH_3. In most cases the complex consists of the central metal atom or ion with up to six Lewis bases clustering around it. In this context, the Lewis bases are known as 'ligands' (from the Latin for 'bound') and I shall use that term here.

I don't want you to think that I am embarking on stratospherically esoteric material again. These metal complexes are hugely important in many aspects of the everyday world. For instance, chlorophyll is a complex of magnesium and is responsible for capturing the energy of the Sun for photosynthesis (Reaction 26). There is hardly a more important molecule. One that comes close in importance is haemoglobin, an elaborate complex of iron, which ensures that oxygen reaches all your cells and keeps you alive. Many pigments are complexes, so your life is decorated and made more colourful by them. Some pharmaceuticals are complexes based on platinum, so one day, perhaps even now, you might be kept alive by one of these artificial complexes. Some catalysts are complexes, so the wheels of chemical industry and all that springs from it are kept turning by complexes.

As an example of a complex and the type of reaction I shall talk about, let's consider a solution of copper sulfate in water. Its delightful pale blue colour is due to the presence of a Cu^{2+} ion, the Lewis acid, at the centre of a complex with five H_2O molecules, the ligands, arranged around it, I. Each ligand molecule uses one of its lone pairs (the two-electron-rich region of a molecule, as I described in Reaction 9) to form a bond that keeps it attached to the metal ion. When a few drops of a solution of ammonia in water are added, the solution turns a deeply satisfying dark inky blue. This colour change is a result of the replacement of some of the H_2O ligands by NH_3 molecules.

Many transition metal complexes behave similarly but with different variations of colour. Very often these changes of colour are accompanied by changes in other physical properties, particularly

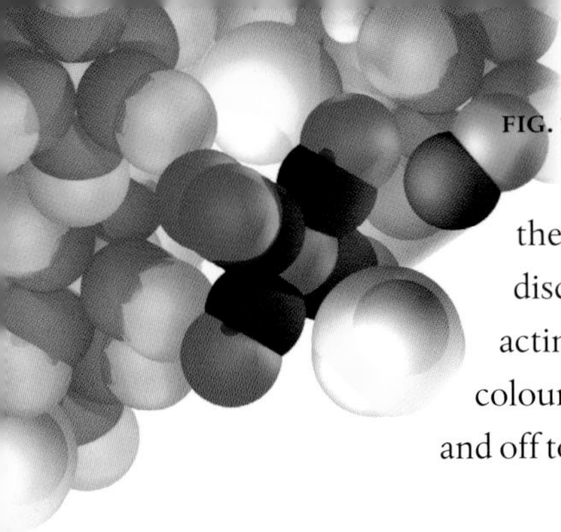

FIG. 10.1 *Departure then attack*

their magnetic properties. The reactions that I discuss in this section are therefore capable of acting as chemical switches not only for turning colours on and off, but for turning magnetism on and off too.

Marriage, separation, and divorce

The switching that I have just mentioned is a result of 'substitution', the replacement of one type of ligand by another. There are several ways in which the substitution might take place. Let's do our usual trick and shrink to the size of a molecule, stand in the midst of solutions where the various types of reactions are taking place, and see what is going on. I shall use different examples in each case because each sequence of steps depends on the identity of the reagent, with different complexes having different chemical personalities and therefore behaving differently.

In one case we see a ligand simply falling off the central atom and drifting away into the surrounding water (Figure 10.1). Its departure leaves the remaining complex open to attack. Quite soon we see the incoming ligand snooping around, detecting the vacated location on the atom, and seizing its chance. It uses its lone pair of electrons (remember that it is a Lewis base) to attach to the atom. We have witnessed the formation of a new complex by substitution.

In another case we see something quite different. As we watch, nothing seems to happen to the initial complex, which survives the buffeting of the water molecules around it. But that changes when another potential attacking ligand wanders by like a shark in the water. We see the incoming attacker approach the complex, and start to form a bond to the central metal atom (Figure 10.2). At the same time, like weaklings everywhere,

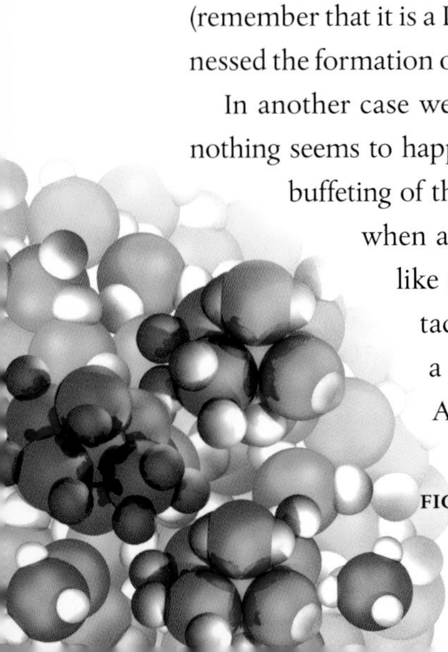

FIG. 10.2 *Retreat in the face of attack*

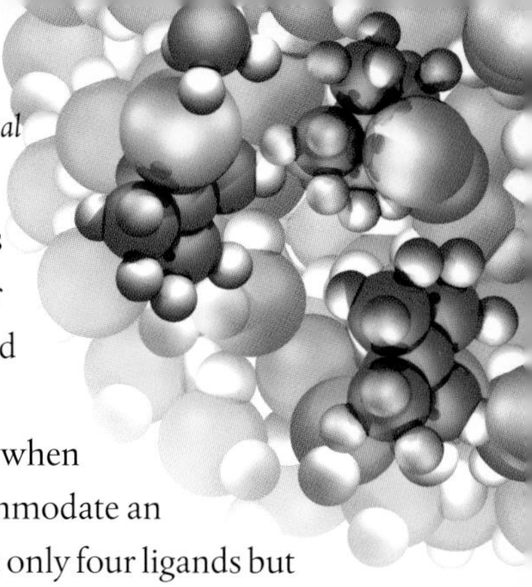

FIG. 10.3 *Acceptance, then dismissal*

one of the ligands seems to accept defeat, and its bond to the metal atom lengthens. The incomer presses home its attack, and the weakling ligand drifts off defeated.

There is a third possibility which is common when the initial complex has plenty of room to accommodate an extra ligand, for instance, when the complex has only four ligands but has room for more. When we watch in this case, we see the ligand destined to be displaced stubbornly resisting (Figure 10.3). The attacking ligand moves in on the complex, but the outgoing ligand initially hangs on to the metal atom for dear life. Then, once the incoming ligand is firmly attached, we see the outgoing ligand accept defeat. It let's slip its hold on the atom, falls off the complex, and slinks away into the metaphorical sunset.

Colour prejudice

I have remarked that reactions like these commonly result in changes of colour. What colour is observed for a complex depends on the manner in which electrons are arranged on the central metal atom. I won't go into this fairly esoteric field in detail, except to say that different ligands result in different energy levels for the electrons on the central atom. Then, because light of different frequencies is able to excite the electrons between the available energy levels, different colours are absorbed from incident white light, leaving the 'complementary' colour to be observed. Thus, if green light is absorbed from white light, then the complex looks red. If a substitution reaction results in different allowed energies such that the complex now absorbs yellow light, then the complex turns violet.

The source of colour I have just mentioned arises from the rearrangement of electrons on the central metal atom itself. Many of the bright colours of complexes, however, arise from intense absorptions in which an electron is boosted from one of the ligands onto the central atom. The resulting colour depends on the identities of both the ligands and the central metal atom. Changing the ligand can therefore affect the colour profoundly. Now you can begin to see—begin, no more than that—how chemists can conjure colours into existence and tune them to their needs.

11

Marriage Broking

CATALYSIS

A 'catalyst' is a substance that increases the rate of a chemical reaction without itself being consumed. The Chinese characters for catalyst, which translate as 'marriage broker', convey the sense exactly. For instance, a huge advance in industrial chemistry was achieved early in the twentieth century when the German chemist Fritz Haber (1868–1934) found that nitrogen and hydrogen could be induced to combine to form ammonia, NH_3, if the two gases were heated under pressure in the presence of iron. They hardly react at all if iron is not present. Haber's achievement has helped to save the world, as well as contributing not a little to its destruction. Ammonia is of prime importance for the production of fertilizers, and through that application catalysis has helped to feed the world. Ammonia is also of prime importance for the manufacture of explosives, and through that application catalysis has taken away with that hand some of what the other hand has provided.

The chemical industry could not function without catalysts as they enable reactions to occur at economically viable rates. They also enable some reactions to occur which in their absence would not occur at all. Catalysts are used to refine fuels, thus enabling transport. They are used in the manufacture of polymers, thus enabling the fabrication of so many of the artefacts of everyday life as well as the fabrics of fashion and furnishings. Without catalysts there would be very little of what we recognize as the familiar modern world. Our bodies also function under the control of catalysts. Biological catalysts are called enzymes, and I describe their function in Reaction 27.

There are two broad classes of catalyst. A 'heterogeneous catalyst' is typically a solid and the reagents are liquids or gases that flow over the solid and react as they come into contact with it; this is the case with Haber's catalyst. A 'homogeneous catalyst' is a gas or a substance that dissolves in a liquid reaction mixture. Anthropogenic (human-made) chorine atoms, perhaps from aerosol gases that have travelled up into the stratosphere, are homogeneous catalysts for the destruction of ozone. I shall postpone my discussion of catalysis by acids and bases until Reactions 17 and 18 when you will see that they play an important role in the transformations of organic molecules.

There is no 'universal catalyst', no philosopher's stone, no elixir that accelerates whatever it touches, and for many reactions there are no known catalysts. Although some catalysts work for a range of reactions, it is usually the case that the catalyst for each reaction needs to be developed individually. There is, however, a common feature that accounts for the action of all catalysts: they provide a different mechanism for the reaction than that allowed when no catalyst is present. Provided the new mechanism is more facile than the old one, then the products are produced more rapidly. The action of a catalyst is like building a new highway to replace a country lane.

REACTIONS

Street life

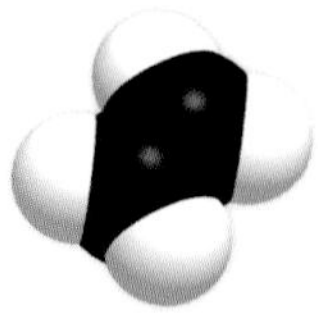

I shall describe an example of heterogeneous catalysis first. Let's suppose you want to use hydrogen to react with ethene, $CH_2=CH_2$ (ethylene, 1), to produce ethane, CH_3-CH_3, 2. There are more useful things to do with ethene than turning it into rather useless ethane, such as using it to make polyethylene (Reaction 13), but this example avoids a lot of complexity. It will also help you appreciate the more useful catalysed reactions that go on in oil refineries where hydrocarbons are 'cracked', broken up into fragments, and 'reformed', their atoms juggled around into new arrangements, into valuable tailored fuels. Just mixing hydrogen and ethene gases and heating the mixture results in no reaction. However, if the mixture is heated in the presence of the metal nickel, Ni, reaction is fast.

Let's shrink to the size of a molecule and take a close look at what is going on at the surface of the nickel (Figure 11.1). The nickel looks a bit like a cobbled street, but there are cliffs several atoms high and all manner of outcrops of atoms. Standing there, we find ourselves in the midst of a storm of collisions from the overlying gas molecules. On many of the impacts, we see the molecules simply bouncing off, but in some cases they form a bond to Ni atoms and lie there tethered to the surface. We see that an ethene molecule has formed two weak carbon–nickel bonds using the electrons present in its double bond.

When an H_2 molecule strikes and attaches to the surface we see that as it forms nickel–hydrogen bonds the H–H bond weakens, the molecule falls apart, and the H atoms become free to move independently. These atoms skid around over the surface. Very soon we see that one collides with an ethene molecule

FIG. 11.1 *Conversion of ethene to ethane*

and forms a carbon–hydrogen bond. At this point the ethene molecule has become CH_2–CH_3, with the C atom of the CH_2 group tethered to a Ni atom and the attached CH_3 group waving around like a flag over the surface.

Then, almost immediately, we see another skidding H atom collide with the tethered molecule. It severs the remaining nickel–carbon bond, and the molecule is released as CH_3–CH_3, ethane.

Street cleaners

The 'catalytic converters' of vehicles are highly sophisticated catalysts that are designed to ensure that the exhaust gases from the engine are reasonably benign and lack poisonous carbon monoxide and the nitrogen oxides that contribute to acid rain. They also need to ensure that unburned hydrocarbon fuel is fully oxidized and not released as rogue molecular fragments into the atmosphere where they can contribute to the formation of photochemical smog (Reaction 24).

In a 'three-way' converter, the catalyst has three duties to guard the environment: to reduce nitrogen oxides to harmless nitrogen, to oxidize unburned fuel to carbon dioxide and water, and to oxidize carbon monoxide to carbon dioxide. This last action has rendered much less reliable the once popular mode of suicide by inhaling exhaust gases. It has not, of course, reduced the impact of transport on the delicate composition of the atmosphere. Platinum and rhodium are used for the reduction stage, and platinum and palladium are used for the two oxidation stages. One problem for designers is the need to find formulations that act even at the early stages of a journey, while the engine is cool. Another is to find measures that prevent thieves from stealing the converters for the precious metals they contain.

3

Let's shrink, as usual, so that we can crawl inside the converter and look at the reduction stage. In this step, nitrogen oxides—which include NO, 3,

FIG. 11.2 *Nitrogen oxides removal*

and NO_2, 4, and are collectively denoted NO_X, with a very faint and perhaps unintended nod towards obno$_X$ious—are converted to benign nitrogen molecules, N_2. There is a hefty gale blowing in the exhaust, with mol-

ecules of many noxious compounds and great boulders of particulate matter hurtling past us, but we see some NO_X molecules of both kinds becoming attached to the surface of the metal catalyst (Figure 11.2). This catalyst is spread as a fine coating on a ceramic honeycomb, but we can see only a tiny patch of it in the murk of the exhaust. We do see, though, that when the NO_X molecules attach to the surface the atoms are dragged apart and the N and O atoms become spread out over it. At the high temperatures involved, these atoms can skip from metal atom to metal atom. Soon two N atoms collide, form a strongly bonded N_2 molecule, and leave the surface. We also catch sight of two O atoms meeting as they hop around; they form O_2 and fly off from the surface.

Similar events take place in the oxidizing regions of the converter, with extra oxygen supplied to bring about the oxidation. If we manage to crawl there, or simply get blown there, we can watch the action taking place on the surface of the metal catalyst (Figure 11.3). We see CO molecules tumbling in and striking the surface. A feature we note is that a CO molecule is able to attach to the surface in several different ways. Some we see standing head on, with the C atom attached to a single metal atom. Another is attached like a bridge spanning two metal atoms. A third has formed bond to a triangle of neighbouring metal atoms. We see that there are also O atoms attached to the surface where O_2 molecules have struck it, have been captured, and have been ripped apart into atoms. As we watch, these O atoms skitter across the surface,

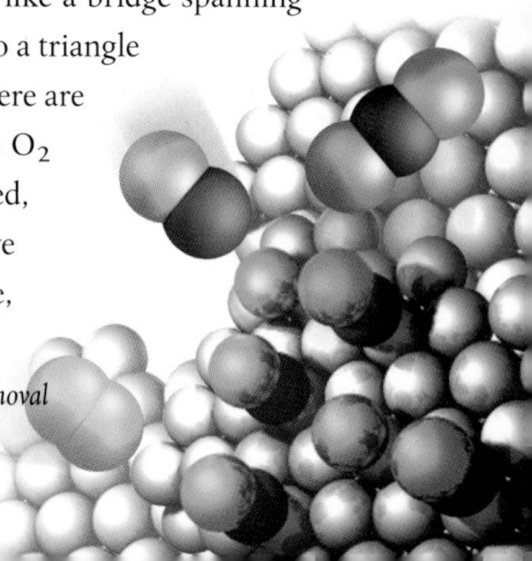

FIG. 11.3 *Carbon monoxide removal*

encounter a CO molecule, form a new carbon–oxygen bond, and leave the surface as much less toxic but still environmentally deleterious CO_2.

As you might appreciate, the worst that can happen to a catalytic converter, apart from its theft, is for the all-important metal surface to become irreversibly contaminated. That is one reason why leaded petrol is no longer used, for when the lead tetraethyl it contained decomposed into ethyl radicals (Reaction 13) and lead atoms, the latter stuck irreversibly to the metal and sealed it against further action.

Aerial combat

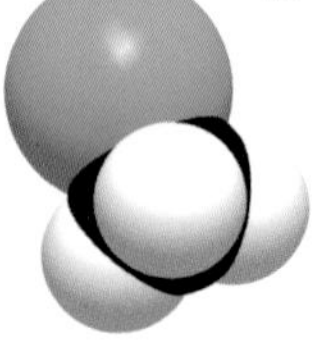

As an example of homogeneous catalysis, I shall describe the havoc that a chlorine atom, Cl, can bring about in the upper atmosphere. Chlorine atoms are produced by the impact of solar radiation on methyl chloride, CH_3Cl, **5**, once widely used as a propellant in aerosol cans and which drifted high into the atmosphere after its propellant duties were done. Methyl chloride is also formed as a by-product of reactions between chloride ions, Cl⁻, and decaying vegetation in salty oceans,

After the murky hot gloom of the interior of the catalytic converter, it is a relief to rise high in the atmosphere to see what is happening there among the ozone molecules of the stratosphere (Figure 11.4). We keep our eyes on a CH_3Cl molecule that has drifted up too. Under the impact of sunlight, we see the molecule shake off a chlorine atom. Very soon we see that atom collide with an ozone molecule, O_3, **6**, and carry off one of the O atoms as ClO, leaving behind ordinary oxygen, O_2. High in the atmosphere there is plenty of atomic oxy-

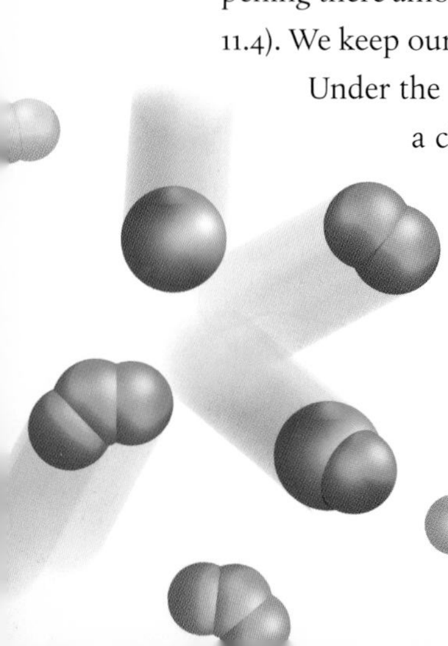

FIG. 11.4 *Destruction of ozone*

gen formed by the impact of solar radiation on both O_2 and O_3 molecules. As we watch we see the ClO molecule collide with an O atom. The O atom plucks the O atom out of the ClO molecule and flies off as O_2. But that hijacking has left a lone Cl atom. It goes on to destroy another O_3 molecule before it is regenerated and is able go on again and again until some other event removes it from circulation. A single Cl atom can thus destroy millions of ozone molecules. It is for this reason that current international protocols prohibit the use of aerosol propellants and refrigerator fluids that, as vapours, are likely to rise high into the atmosphere and act as a source of Cl atoms.

Divorce and Reconciliation

RADICAL RECOMBINATION

In Reaction 3 you saw that a radical is a species with at least one unpaired electron. An 'unpaired electron' is a single electron that is present in the molecule but not playing a role in bonding. The French word *celebataire* conveys the sense of the electron's forlorn loneliness very well. However, it is capable of joining forces with another unpaired electron on another radical to form a bond. Two examples of radicals are ·OH (1, a hydroxyl radical), and ·CH$_3$ (2, a methyl radical). The dot denotes the unpaired electron. In most cases, radicals are highly reactive and aggressively attack other species in order to use their un- paired electron to pair with an electron on the second species and so form a bond. The most primitive type of radical reaction is simply the clunking together of two radicals, each donating its unpaired electron to the formation of an electron-pair bond, as in the combination of two ·CH₃ radicals to form ethane, CH₃–CH₃, 3.

Some species might have more than one un-paired electron that they can use for biting into other molecules. If they are double-fanged, the most common case after ordinary single-fanged radicals, then the species is known as a 'biradical'. To continue the partnering analogy: the two electrons cohabit but are platonic in their relationship. A very important example is an oxygen atom, O, which is a biradical. For the purposes of this section I shall write it ·O· with two dots for its two relevant electrons. An O_2 molecule is also a sort of biradical, so when I want to emphasize its radical nature I shall denote it ·O–O·.

Pedant's point. For certain technical reasons to do with the way that the spinning directions of the two unpaired electrons are aligned with each other, ·O· is not a true biradical, but we can treat it as one for our purposes. The same is true of O_2.

There are a lot of reasons why you should be interested in radical reactions. One is that, as I explained in Reaction 3, they take part in the combustion reactions of the everyday world. Combustion reactions include the reactions that take place inside internal combustion and jet engines and move us around the world. Radical reactions occur wherever there is fire. Because fire is sometimes unintended, if the reactions that contribute to it are understood, then better ways to control and quench it can be devised. Radical reactions are also involved constructively in some of the processes that contribute to the modern world, for they are used in the manufacture of certain kinds of plastics (Reaction 13). They are also of huge importance in the upper atmosphere, where they maintain its composition and help to protect us from the dangerous vigour of the Sun's radiation. Radicals are also involved in biological cell degeneration and its regrettable manifestation, aging. We can but hope that an understanding of radical reactions might in future provide the key not only to combustion but also to something approaching eternal youth.

So how are radicals made? Sunlight can rip molecules apart, and so can its laboratory emulation in the form of ultraviolet radiation. A match and a spark can do the same job, as you saw in Reaction 3. But what can a chemist do to make radicals under carefully controlled conditions? One procedure is to

add to a reaction mixture a small amount of a compound that has molecules with a very weak bond and then shake those molecule into fragments by heating it. Heating causes all the molecules in a mixture to jostle each other vigorously, and the two parts of a flimsy molecule may be knocked apart. Molecules with an O–O bond, which are called 'peroxides', are quite flimsy and are commonly used. For instance, $CH_3O–OCH_3$, 4, has an exceptionally feeble O–O bond and the molecule falls apart into two $CH_3O\cdot$ radicals, 5, when heated or exposed to light. The compound 6, which will figure in Reaction 13, behaves similarly, falling apart into halves quite easily when it is heated. The compound 7 behaves differently but with

a similar result. When it is heated, the N=N group is blasted out of the molecule as an N_2 molecule, leaving two radicals.

An important aspect of radical reactions, which I will make clear in the course of this section, is that they commonly occur as a chain, with one event as a link that leads to another. In such a 'chain reaction', one radical attacks a molecule, does its business, such as extracting an

H· atom, and converts the victim molecule into a radical. That radical attacks another molecule, so producing yet another radical, and so on. The chain of reactions continues until it terminates in some way, such as by the encounter of two radicals, which snap together without forming a new radical and bring the chain of links to an end.

Making water

I shall introduce chain reactions by talking about one that at first sight appears to be simple: the reaction of hydrogen, H_2, and oxygen, O_2, to give water, H_2O. To some extent, this reaction is similar to the combustion of a hydrocarbon, which is also a chain reaction and which I described in Reaction 3. It introduces several interesting new aspects of reactions that involve radicals, in particular the possibility that the reaction mixture might explode. This is the mechanism behind the characteristic 'pop' when a mixture of hydrogen and oxygen is ignited with a spark or flame.

To see what is going on, we imagine that we are standing in the mixture of hydrogen and oxygen, with H_2 and O_2 molecules hurtling past us and colliding with one another. Then someone applies a spark. Suddenly, we find ourselves in a furious storm of events, with molecules being ripped apart into atoms and what seems like lightning bolts flashing through sky but in fact are pulses of light being given off as electrons blasted off atoms settle back down onto them and release energy as packets of electromagnetic radiation. As the storm subsides, we are able to focus on the events that it has unleashed.

We see two hydrogen atoms that are formed when an H_2 molecule is torn apart (Figure 12.1). These atoms are radicals, so from now on I write them as H·. As we watch, we see one of them as H· atoms collide with an O_2 molecule. After a brief struggle it carries off an ·O· atom as ·OH, leaving behind a two-

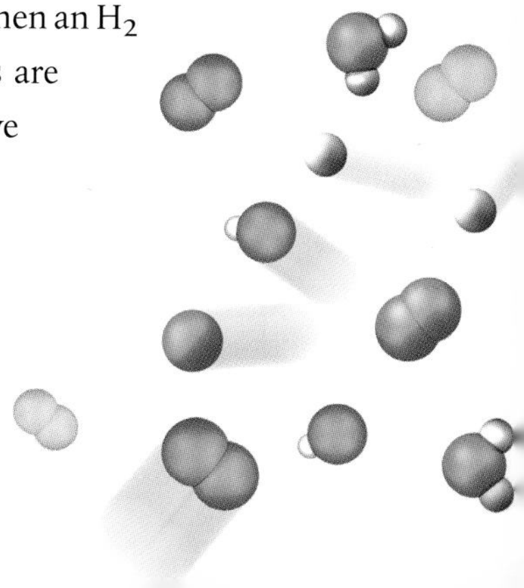

FIG.12.1 *Hydrogen and oxygen forming water*

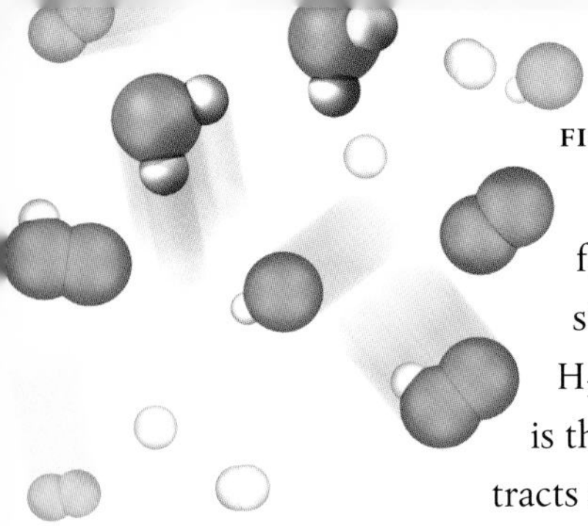

FIG. 12.2 *Termination*

fanged $\cdot$O$\cdot$ atomic radical. We then just catch sight of the newly formed $\cdot$OH radical striking an H_2 molecule that survived the initial storm. There is then another brief struggle. In it, the radical extracts an H$\cdot$ atom to form H_2O, and the remaining H$\cdot$ atom flies off free to react elsewhere.

We also catch sight of an event in which death comes to a chain. As we watch the collision of an H$\cdot$ atom with an $\cdot$OH radical, we see that the two radicals stick together and form an H_2O molecule. Each radical has donated its odd electron to the pair needed for the formation of a new chemical bond. The formation of an H_2O molecule is accompanied by the removal of two radicals from the chain of processes. We catch sight of death of another kind when we see another H$\cdot$ atom collide with an $\cdot$O–O$\cdot$ molecule and forms HO–O$\cdot$ (Figure 12.2). In this step, three unpaired electrons, one on H$\cdot$ and two on $\cdot$O–O$\cdot$, have ended up as a bond and only one unpaired electron is left. The collision has effectively replaced three 'radicals' by one.

These reactions, and other like them, go on until all the hydrogen or all the oxygen has been converted into water.

There is a sneaky aspect of this account that you might have noticed. One event we see is the collision of an $\cdot$O$\cdot$ atom with an H_2 molecule with the formation of $\cdot$OH and H$\cdot$ radicals. Now, although there remain just two unpaired electrons, those electrons are carried by two radicals not the single $\cdot$O$\cdot$ atom. This separation gives them much greater mobility. Moreover, because each radical can start up a new chain of reactions, you can expect there to be a surge in the reaction rate. In fact this so-called 'branching' step can initiate chains that themselves branch in a similar way, and suddenly the reaction becomes very fast. In other words, there is an explosion and the sparked mixture gives a characteristic 'pop'.

Knock for knock

The elimination of dangerous radicals—dangerous in the sense that they might lead to branching and therefore unwanted explosion—was the strategy behind the once widespread addition

of lead tetraethyl, 8, to gasoline, for the Pb atom readily sheds its ethyl group, $CH_3CH_2\cdot$, in the hot tumult of the ignited fuel–air mixture, and these radicals can hinder the propagation of branched chains and hence lead to smooth, efficient combustion rather than explosion and 'knocking'. With lead banned, partly for its environmental and health impact but also because it would poison the catalytic converters in cars (Reaction 11), refineries have looked for more benign sources of these scavenging radicals, and some use $CH_3OC(CH_3)_3$, 9, which also falls apart into radicals reasonably readily.

Flame retarding fabrics are based on a similar antiknock strategy. If bromine compounds are incorporated into a fabric, then if it catches fire the bromine atoms, Br·, are released. These atoms are radicals, and can combine with other radicals that are propagating the chain that corresponds to combustion, and it is extinguished.

II

ASSEMBLING THE WORKSHOP

The rest of the reactions in this book focus on the ways in which intricate molecules based around carbon—the organic molecules—are constructed. Here you will see how the tools I have assembled for you so far can be used to build molecules of almost unlimited intricacy and thereby enrich the world with substances of astonishing usefulness, substances that perhaps do not exist anywhere else in the universe.

Almost all the reactions so far have been blunderbuss affairs, in which the encounter of reagents collapses into some kind of product, showing little subtlety of interaction. These reactions have been like unloading a pile of bricks and hoping that they will tumble into the form of a cathedral. The reactions I shall now describe are far more subtle. In them, one atom is coaxed to attach to another and the process is more like the careful crafting of a cathedral, stone by stone, and decorating it with exactly the right ornaments. Organic chemists, the chemists whose principal role is the almost magical pursuit of intricate structures like the one shown in the underlying image here, have developed an extraordinary array of subtle techniques for achieving this coaxing, and all I can do in the following pages is to illustrate a tiny fraction of them.

I need to make you aware of the following remark, which underlies how our tools are applied to build intricacy. There is an underlying distinction between what has gone before and what is to follow, and which essentially accounts for the change from blunderbuss to stiletto. In the foregoing blunderbuss reactions the guiding principle has been the collapse of a system into lowest energy. They take place as though a trapdoor is opened beneath the feet of the reactants, which then fall through to form the products below.

> *Pedant's point.* To be more precise, they are driven by the thermodynamic tendency to achieve lower Gibbs energy, a quantity that takes into account the tendency to achieve greater entropy.

In the stiletto-like reactions of organic chemistry, in contrast, there is not so much a trapdoor to drop through to achieve lower energy but a network of openings and corridors to clamber down. If the reagents can be encouraged to pass through one door rather than another, then they might enter a room from which there is no escape, even though it is not a room of lowest energy. That room might correspond to a compound that can be used as a reagent in another reaction. If it is so used, then it might enter another trapping room further on in the journey towards the desired final product, the deepest dungeon, the fully decorated cathedral.

> *Pedant's point.* In this analogy I am trying to convey the distinction between thermodynamic control ('blunderbuss') and kinetic control ('stiletto').

REACTIONS

13
Stringing Along
RADICAL POLYMERIZATION

Perhaps the most striking observable change brought about by chemistry as applied to the everyday world is in the texture of objects. Until the early twentieth century objects were manufactured from metal, wood, and animal and plant fibres. Today, synthetic polymers, in the vernacular 'plastics', are ubiquitous and have changed not only the appearance of the world but also its feel.

Polymers (from the Greek words for 'many parts') are made by stringing together small molecules, the 'monomers', into long chains or extensive networks. Thus, polyethylene (less formally polythene) is a chain of linked ethylene molecules, and polystyrene is a chain of linked styrene molecules. In some instances, two or more different types of monomer molecules are used to form the polymer. Thus, one form of nylon is a chain in which the alternating links are of two different compounds.

There are two main ways of linking molecules together, one involving radicals (Reaction 12) and the other not. In this section I shall introduce you to radical polymerization and treat the other kind in Reaction 14. Polyethylene and its cousins are made by the radical method, and I start with them.

An ethylene molecule (1, formally, ethene) is written $H_2C=CH_2$, the double bar denoting a 'double bond'. This is the first time I have needed to introduce you to a double bond, but it will turn out to be a crucial feature of all the monomers in this section. A double bond consists of two ordinary bonds linking the same two atoms. Because each bond consists of a shared pair of electrons acting as glue between two atoms, a double bond consists of two such pairs. Although a double bond between two C atoms is stronger than a single bond between C atoms, it is not twice as strong because the two pairs of electrons struggle for the best position and tend to push each other out of the ideal location for bonding. One consequence is that a molecule can acquire lower energy by giving up one of the pairs of electrons in the double bond and forming more single bonds with other atoms. In other words, a carbon–carbon double bond is a reaction hot spot in a molecule, and you should expect it to be the site of a lot of chemical action.

The four H atoms of ethylene play no role in what follows and just come along for the ride. You might therefore be able to guess that one or more of these atoms can be replaced by other atoms or groups of atoms and thus give rise to all manner of different versions of polyethylene. Indeed, that is the case. It is the origin of polystyrene, in which one of the H atoms has been replaced by a benzene ring, 2. It is also

the origin of a lot of other derivatives, such as PVC (polyvinylchloride) in which one H atom has been replaced by a chlorine atom, Cl, and polytetrafluoroethylene (PTFE, sold as Teflon), in which all four H atoms are replaced by fluorine atoms, F.

FIG. 13.1 *Ethylene starting to polymerize*

Hairy spaghetti

A radical reaction needs to be initiated. One straightforward procedure is to introduce a molecule that has a bond that readily falls apart into two radicals when heated. I mentioned some examples of such fragile molecules in Reaction 12. One common initiator molecule

with a sickly O–O bond is **3**, which obligingly falls apart into two radicals, **4**, when it is heated and jostled by other molecules.

Let's shrink, as usual, and go to the heart of the action and watch what is going on when one of these initiator radicals has been formed in otherwise pure, compressed ethylene. As always in a gas, there is a whirlwind of activity as the ethylene molecules hurtle past, collide, and ricochet off one another. In one of the collisions near us we see an ethylene molecule collide with an initiator radical (Figure 13.1). That radical sinks its single tooth into the

electron-fat double bond of the ethylene molecule. As it does so, it forces one electron pair of the bond to separate into individual electrons. One of those newly and only briefly celibate electrons immediately pairs with the radical's single electron, so forming a bond between the radical and the molecule. The other electron of the original pair is left stranded on the other C atom of the ethylene molecule, so ethylene has become a radical with some initiator baggage.

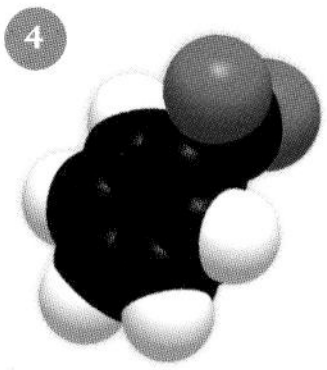

We now see this big radical colliding with another nearby ethylene molecule that zooms in on its unwary flight (Figure 13.2). In much the same way, one component of the latter's double bond springs open, a bond forms to one of the C atoms, and an electron gets stranded on its other C atom of the newly captured ethylene molecule. At this point, the gas contains an

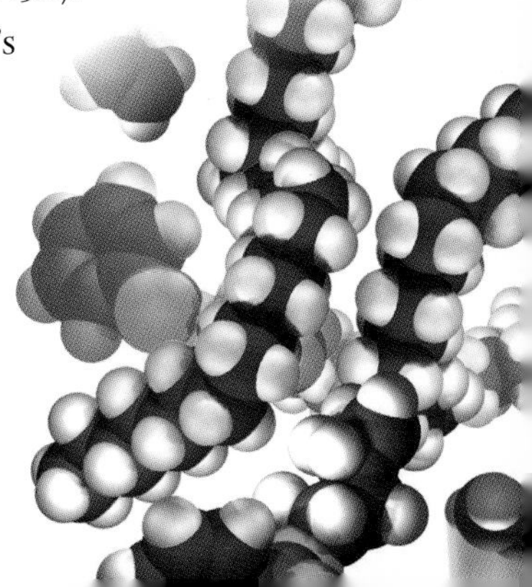

FIG. 13.2 *Chain propagation*

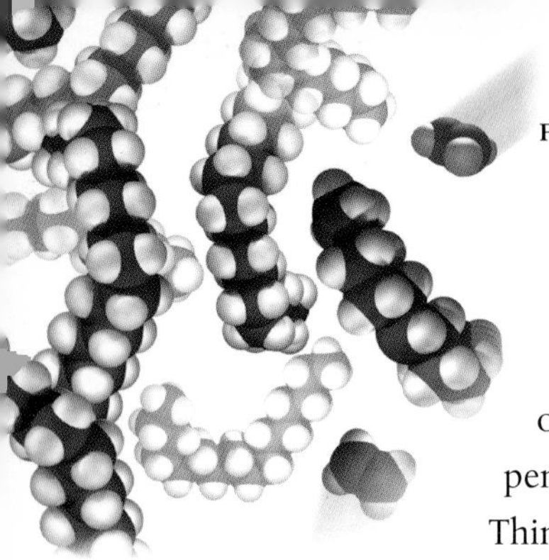

FIG. 13.3 *Cross-linking and termination*

even heftier radical. As we continue to watch we see the chain growing link by link at each successive collision between the business end of the radical and an ethylene molecule that happens to strike it there.

Things can go wrong. In the distance we see the business ends of two lengthy chains meet (Figure 13.3). The electrons on each then pair and form a bond. As a result, a lengthy polymer molecule forms, but at the expense of terminating the chain reaction. Elsewhere we catch sight of a different kind of attack. The business end of a longish, writhing chain strikes somewhere along another growing, writhing chain. There is a brief shudder as the electrons in a carbon–hydrogen bond reorganize themselves in the presence of the incoming electron, and as the second molecule snaps away we see that it has carried off an H atom from the first molecule. There are two consequences.

One is that the second radical, by carrying off the H atom, has become an ordinary non-radical molecule and can no longer participate in the chain. The other is that the first molecule has been left with an unpaired electron on a C atom somewhere along the chain as well as at its untouched end. The chain now has a business middle as well as a business end. As we watch, we see an ethylene molecule collide at the mid-chain hot spot, form a bond, and start to grow a side-chain.

As can be surmised from this account, the product of radical polymerization may be a collection of many different chain lengths and cross-linkages and side chains. The polyethylene molecules that formed are becoming more like a hairy version of spaghetti.

Bald spaghetti

A much more controlled version of the polymerization minimizes the hair. As a result, the molecules are much more like ordinary

FIG. 13.4 *Ziegler–Natta catalyst*

spaghetti, bald spaghetti, and can pack together much more efficiently, giving a denser material. The key is a catalyst discovered by a German chemist, Karl Ziegler, and an Italian chemist, Giulio Natta, in the 1950s; it earned them the Nobel Prize in 1963.

The catalyst is based on a titanium chloride and an organic compound based on aluminium, Al. When you look closely at the surface of the solid (Figure 13.4), you see that a CH_3CH_2 group has transferred from an Al atom to a titanium atom, Ti. Crucially, the Ti atom also has an exposed face, a site capable of forming a bond but as yet unoccupied by other atoms.

Now watch what happens when ethylene molecules shower on to the surface (Figure 13.5). Where an ethylene molecule happens to hit the exposed Ti atom, it attaches. There is then a little electronic skirmish. When the metaphorical dust settles we see that a CH_3CH_2 group has migrated onto the attached $CH_2=CH_2$ molecule. That rearrangement has left a vacant bonding position where the CH_3CH_2 group was initially. As we watch, another $CH_2=CH_2$ molecule strikes the Ti atom and attaches to it. The same sort of electron skirmish then recurs and the outcome is as lightly longer chain anchored to the solid. As we continue to watch, we see a growing chain spin out from the Ti atom on the surface of the catalyst like a growing spider thread that in due course can be harvested as long molecules of polymer with very few side chains and cross links.

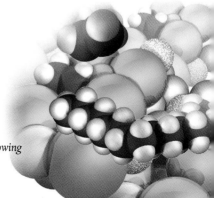

FIG. 13.5 *Polyethylene growing*

14
Snapping Together
CONDENSATION
POLYMERIZATION

One of the most famous of all plastics is nylon. I shall use it to represent how the second type of polymers is made. There are many varieties of nylon, but it will be enough to consider just one exemplar, the one known as 'nylon-66'. It is so called because the repeating motif is a chain of six carbon atoms, then a group of atoms that provide a link, and then another chain of six carbon atoms. The pattern C_6-link-C_6 is repeated indefinitely to give the '66' polymer. As you will see in more detail in Reaction 27, nylon is a very primitive version of a protein-like molecule, the molecules that control all bodily process and also form claws, nails, and hair. A protein molecule has the same links but more varied combinations of carbon atoms. It is, however, an interesting point that we clad our exteriors in material that emulates our interiors.

One of the molecules used to build the polymer is 1 (on the next page; note the six C atoms). As it happens, this molecule is a close

relative, with four and five C atoms respectively instead of six, of the two compounds cadaverine and putrescine, with names indicate their origin and odour. Thus, not only does nylon emulate the living, but we clad ourselves in molecules akin to the odour of death. The other molecule we need is 2 (note the six C atoms again). The task of the nylon manufacturer is to link the $-NH_2$ end of one molecule to the $-COOH$ end of the other molecule, then doing it over and over again to grow long spindly polymer molecules. You know from Reaction 2 that NH_3 is a base (a proton acceptor), so you should be able to accept that the $-NH_2$ business end of molecule 1 is also a base. Similarly, you saw in the same discussion that $-COOH$ is a structural motif of acetic acid, an acid (a proton donor), so you should be able to see that the business end of molecule 2 is an acid. Manufacturing nylon, should therefore be seen as starting off as one giant acid–base neutralization reaction with 1 pitted against 2.

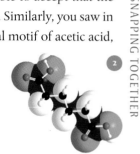

As usual, to see what is going on, we shrink to the size of a molecule and immerse ourselves in the reaction mixture. This is not a gas; we are in a dense forest of molecules mingling together in water (Figure 14.1). All around us there is a ceaselessly writhing and wriggling dense molecular undergrowth and an abundance of water molecules.

When an acid, a proton donor, finds itself in the company of a base, a proton acceptor, a proton hops from the acid to the base. We are standing in a crowd of proton donors and acceptors, and we see this transfer going on wholesale wherever we look. Then, as a result, instead of molecules with $-NH_2$ base ends and $-COOH$ acid ends we find ourselves surrounded by molecules with $-NH_3^+$ ends and $-CO_2^-$ ends. That's about all that happens at

FIG. 14.1 *The monomers mixing*

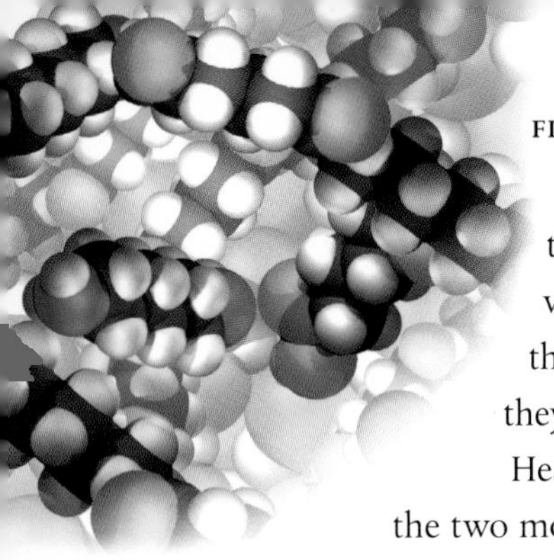

FIG. 14.2 *The monomers reacting*

this stage; it is a start, but at this point there is very little reaction of the kind we want because the molecules have not actually linked together: they are just lying demurely side by side.

Heating comes to our rescue; it causes the ends of the two molecules we are watching to bang together and engage in a little atomic turmoil. One of the protons of the $-NH_3^+$ group escapes back onto the $-CO_2^-$ group of the second molecule, and the exposed dense electron cloud of the N atom of the $-NH_2$ group drives its way into the C atom of the $-COOH$ group (Figure 14.2). The electron cloud of the C–OH bond retreats in the face of this attack and a carbon–nitrogen bond is formed as the OH^- group begins to slink away, pulling a proton off the $-NH_2$ group as it goes and becoming H_2O. All this atomic rearrangement is over in the blink of an atomic eye, and we see the outcome: two molecules are joined by the formation of an $-NH-CO-$ link and a water molecule is expelled to get lost in the crowd. The fusion of two molecules with the loss of a little H_2O molecule is the 'condensation' that gives this type of reaction its name.

We have seen the tentative initial birth of a nylon molecule, with two molecules joined together. Each end of the new molecule has a hot spot like the two that joined forces in the first step, and condensation can occur at each end. We see that at the surviving $-NH_3^+$ end of the first base molecule there is a furious encounter with the $-CO_2^-$ group of another acid molecule, and another $-NH-CO-$ link is formed with the expulsion of an H_2O molecule. Now the chain is three molecules long. Over there, at the other end of the growing chain, there is a similar encounter with the surviving $-CO_2^-$ group of the original acid molecule and the $-NH_3^+$ end of yet another base molecule, and a new link is formed as another H_2O molecule is spat out.

As we stand in this hot, threshing milieu we find ourselves being enwrapped in threads of growing polymers as individual molecules

REACTIONS

fuse on to the active ends of the original molecules and water mol
ecules are squeezed out from between them as they condense together. Unlike in radical polymerization (Reaction 13), where the
formation of side chains is a hazard, nylon chains can grow only at
their ends, not at some intermediate branching points. As a result, the
long hairless chains are well suited for spinning into yarn.

15

Missile Deployment

NUCLEOPHILIC SUBSTITUTION

The equivalent of a heat-seeking missile in chemistry is a 'nucleophile', a lover of nuclei. The 'missile' in this case is a special sort of molecule which sniffs out not heat but positive electric charge, the charge of a nucleus glinting through depleted regions of electron cloud in a target molecule. A candidate for such a missile is a negatively charged ion, which can be attracted to the partially exposed positive charge of a nucleus in the target.

With nucleus-sniffing in mind, you can perhaps appreciate that if it is possible to blow away some of the electron cloud from the region of a particular C atom in the target molecule, then the nucleophile will home in on that atom and perhaps attach to it. By management of the electron clouds and adroit choice of the incoming missile, therefore, it is possible to build where and what we want. You shouldn't lose sight of the fact that the 'missile' doesn't just fly through the air and smash into a molecule: it jostles past the solvent molecules, which move aside or sometimes block it and turn it back. When thinking about the reactions in this section, think of the target atom in a molecule as glowing with positive charge and the incoming missile, the reagent, as having an abundance of negative charge spread over at least one of its atoms. The target and missile are drawn together through the solvent, and might snap together, in a manner I shall describe, once they are close to one another. A missile's successful hunt for a target might then result in reaction and the formation of a new compound.

Target preparation

How are the target prepared and the missile launched? One procedure is to go the whole electrical hog and prepare a molecule with a full positive charge on the C atom that has been chosen to be the foundation of the new construction, the proposed site of reaction. The whole hog can be achieved by snipping out an atom that initially is attached to the C atom. If the extracted atom drags an additional electron with it too, so that it comes away as a negative ion, then it will leave behind a target with a positive charge on the C atom of choice, which is just what is required. A candidate molecule for this kind of procedure is an 'alkyl bromide', a hydrocarbon with a bromine atom, Br, in place of a hydrogen atom, ɪ.

Suppose you want to replace the Br atom with an −OH group or something more ambitious, such as an −OCH$_2$CH$_3$ group or something even more elaborate but of the same general kind. The missile of choice would be an OH⁻ ion (or something similar with H appropriately replaced). It's easy enough to provide OH⁻ ions: just pour in sodium hydroxide, NaOH, dissolved in a mixture of ethanol (ordinary 'alcohol') and water. This solution supplies sodium ions, Na⁺, and hydroxide ions, OH⁻. The Na⁺ ion is not just an idle travelling

FIG. 15.1 *Substitution of the first kind*

companion of the nucleophile: it has work to do too, as you will shortly see.

Now let's watch what is going on in the newly prepared reaction mixture immediately after the hydroxide solution has been poured in (Figure 15.1). You need to bear in mind that a carbon–bromine bond is not very strong. We are immersed in the solvent, and every so often alkyl bromide molecules, Na^+ ions, and OH^- ions wander by. We catch sight of a Na^+ ion drifting across to an alkyl bromide molecule, and see something extraordinary.

When the ion touches the Br atom of the molecule, its positive charge pulls the Br atom out, like a dentist extracting a tooth, as a bromide ion, Br^-. That extraction leaves behind a 'carbonium ion', a C atom and its attachments with a positive charge on the C atom, 2, which is just what is needed for a missile to build on. Once formed, the carbonium ion is ripe for attack by a nearby OH^- ion that senses the positive charge, and we see one bustle across from where it had just been drifting by, attracted by the positive charge. It snaps into place when it is close enough and forms a carbon–oxygen bond.

> *Pedant's point.* This sort of reaction is denoted S_N1. The S denotes 'substitution', the N 'nucleophilic', and the 1 the formation of a single intermediate entity (the carbonium ion), the rate of formation of which controls the rate at which products are formed.

One feature that will play an important role in your future molecule-construction business is that in the particular example I am illustrating the incoming nucleophilic missile can attack the carbonium ion from either more or less flat molecular face. If it attacks from the side originally occupied by the Br atom, the general shape of the

FIG. 15.2 *Umbrella inversion*

molecule is preserved. If you think of the alkyl bromide as an inverted wind-blown umbrella with the Br atom forming the handle, then in this case there is simply a replacement of the handle. However, in half the cases, we see the nucleophile attacking from the side opposite to the one originally occupied by the handle and the inverted 'umbrella' shape is turned inside out as the new handle is formed (Figure 15.2). This turning inside out may be put to good use in certain circumstances as it is one more example of the architectural control that you will be able to exercise over the structure of the molecule you are aiming to build.

Target acquisition

Another type of nucleophilic substitution reaction takes place by the missile, the nucleophile, preparing its own target. There is no prior preparation of a target followed by a strike: preparation and strike occur simultaneously. For simplicity, I shall use the same reagents as earlier in this section, and then explain how to switch between the two mechanisms.

We shrink to see what is happening (Figure 15.3). We are immersed in the reaction mixture again, with the alkyl bromide molecules, Na^+ ions, and the OH^- ions swirling around us in the water. We catch sight of an OH^- ion, the nucleophile, approaching the target molecule from the side that is opposite to the Br atom (from inside the inverted 'umbrella', the Br atom forming the handle). As it approaches, the OH^- ion starts to form a bond to the C atom and the 'umbrella' begins to invert. Simultaneously the bond from the Br atom, the old 'handle', starts to lengthen and weaken. The process continues until the molecule has been turned completely inside out, the incoming OH^- ion has formed a strong bond to the C

FIG. 15.3 *Substitution of the second kind*

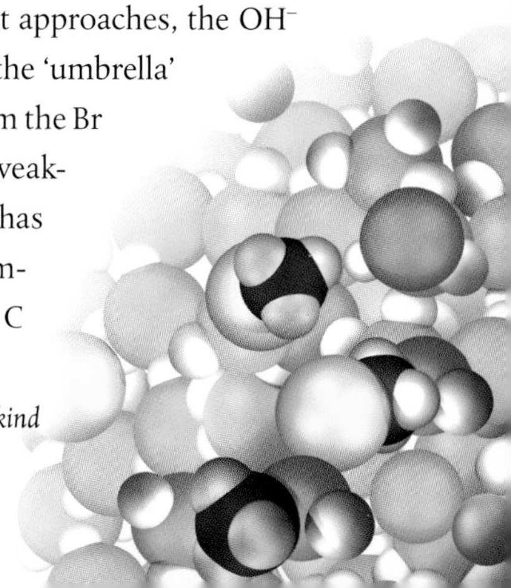

atom, and the Br atom has left completely (as a Br⁻ ion). According to this mechanism, the umbrella must turn inside out, unlike for the first type of reaction I described, where two outcomes are equally likely. You might try to contrive this mechanism when you need a carbon atom to turn inside out in a particular molecular design that you have in mind. Nature sometimes uses it to adjust a molecule to the shape her functions demand.

> *Pedant's point.* Because this kind of nucleophilic substitution reaction involves two reactant molecules simultaneously, chemists denote it S_N2.

Taking shelter

Your building plans might want to avoid this umbrella-like inversion. Fortunately, there are clever ways of ensuring that the umbrella doesn't change its general shape even though the reaction proceeds in this way. Here is just one example. Suppose that one of the H atoms of an alkyl chloride (which is like an alkyl bromide but with a Cl atom in place of the Br atom) has been replaced by a longish flexible group of atoms, as in 3. Let's watch what happens when OH⁻ ions are added to a solution of this molecule (Figure 15.4).

The first action we notice is quite different from what we have seen before in this section. We see a proton transfer, of the kind familiar from Reaction 2, in which the proton migrates from the OH group of the side chain of the molecule onto a nearby OH⁻ ion. That creates an H_2O molecule, which wanders off and gets lost in the crowd.

Next, we see something akin to self-mutilation. The side chain swings round and the O

FIG. 15.4 *Protection against inversion*

atom (which is now negatively charged because it has lost the attached proton) attacks the C atom from the rear. That attack expels the chlorine Cl atom as a Cl⁻ ion and pulls the umbrella inside out. The new arrangement of atoms as a triangular ring is reasonably stable and the self-mutilated, electrically neutral molecule wanders off into the rest of the mixture. But it doesn't survive for long in the dangerous world it inhabits, because it soon encounters an OH⁻ ion. The O atom attached to the C atom has sucked away some of the electron cloud around the C atom, so that the nuclear charge shines through. As a result, the C atom now has a partial positive charge that the OH⁻ nucleophile will sniff out. The ion plunges in, forms an incipient bond and drives out the O atom on the other side. The umbrella turns inside out again. Because overall the umbrella has been inverted twice, it ends up in its original wind-blown arrangement.

Target practice

When should you expect each type of reaction? For the first type of reaction to take place, the carbonium ion needs to be reasonably stable so that it can both form and survive long enough in solution for reaction to take place. It would also help if the C atom at the focus of the attack has attached to it several big, lumpy groups that hinder the approach of the incoming missile. Then that atom is hunkered down and the incipient formation of an intermediate with five groups attached to the target C atom is difficult because the groups already present don't leave much shoulder room. On the other hand, you can expect the second type of reaction when the groups attached to the target atom are small so that the five-group intermediate can form readily.

The first type of reaction is also favoured if the medium encourages the formation of an ion. Ion formation is encouraged if the solvent molecules can cluster around the ion and interact with it favourably. They can do that if they have partial electric charges, with their negative partial charge able to be attracted to the positive charge of the ion. That is the case with solvents like methanol and ethanol, with the dense electron cloud and accompanying negative charge on the O atoms of the molecules. If solvent molecules without such partial charges are substituted, then there is no favourable interaction and the carbonium ion is unlikely to form. You are beginning to see the delicacy of the control that chemists are able to exert as they plan their molecular constructions.

16
Electronic Warfare

ELECTROPHILIC
SUBSTITUTION

Benzene, I, is a hard nut to crack. The hexagonal ring of carbon atoms each with one hydrogen atom attached has a much greater stability than its electronic structure, with an alternation of double and single carbon–carbon bonds, might suggest. But for reasons fully understood by chemists, that very alternation, corresponding to a continuous stabilizing cloud of electrons all around the ring, endows the hexagon with great stability and the ring persists unchanged through many reactions. The groups of atoms attached to the ring, though, may come and go, and the reaction type responsible for replacing them is commonly 'electrophilic substitution'. Whereas the missiles of Reaction 15 sniff out nuclei by responding to their positive electric charge shining through depleted regions of electron clouds, electrophiles, electron lovers, are missiles that

I

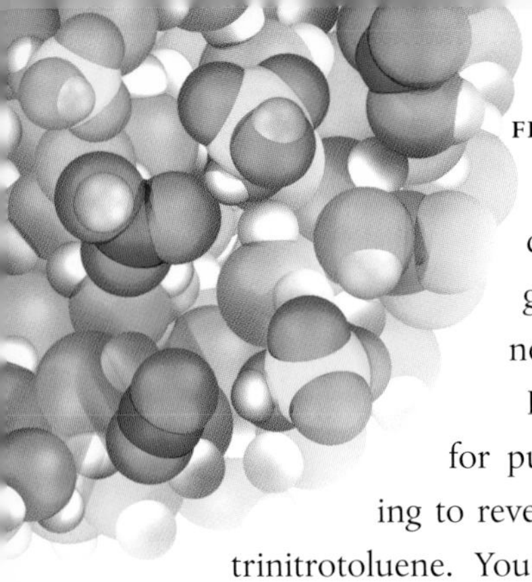

FIG. 16.1 *Preparing the agent*

do the opposite. They sniff out the denser regions of electron clouds by responding to their
negative charge.

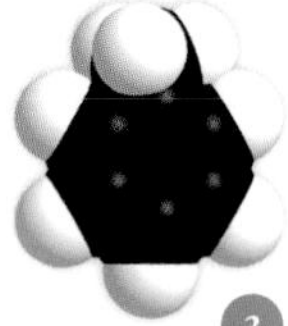

Let's suppose you want to make,
for purposes you are perhaps unwill-
ing to reveal, some TNT; the initials denote
trinitrotoluene. You could start with the common
material toluene, which is a benzene ring with a methyl group
($-CH_3$) in place of one H atom, 2. Your task is to replace three of the
remaining ring H atoms with nitro groups, $-NO_2$, to achieve 3. And
not just any of the H atoms: you need the molecule to have a symmetrical array of these groups because other arrangements are less
stable and therefore dangerous. It is known that a
mixture of concentrated nitric and sulfuric acids
contains the species called the 'nitronium ion',
NO_2^+, 4, and this is the reagent you will use.

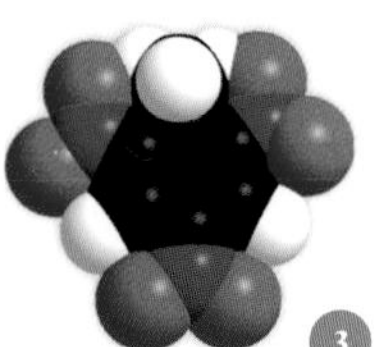

Before we watch the reaction itself, it is instruc-
tive to see what happens when concentrated sulfuric acid and nitric
acid are mixed. If we stand, suitably protected, in the mixture (Figure
16.1), we see a sulfuric acid molecule, H_2SO_4, thrust a proton onto a
neighbouring nitric acid molecule, HNO_3. (Funnily enough, according to the discussion in Reaction 2, nitric 'acid' is now acting as a base,
a proton acceptor! I warned you of strange fish in deep waters.) The
initial outcome of this transfer is unstable; it spits out an H_2O molecule which wanders off into the crowd. We see the result:
the formation of a nitronium ion, the agent of nitration
and the species that carries out the reaction for you.

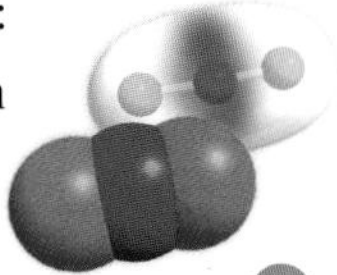

Priming

The positive charge of the nitronium ion lets it function as a seeker
of dense electron cloud and its associated negative charge. A ben-

zene molecule itself has the same reasonably rich density of electron cloud on each of its six C atoms, but when certain groups of atoms are attached to the ring that uniformity is disturbed. Methyl is one such group. When it is present, as it is in toluene, it increases the total electron density in the ring. It also causes the electron cloud to accumulate slightly on alternate C atoms, starting with the immediate neighbours of the C atom to which it is attached. So, when we picture a toluene molecule, we should think of it as having hot spots for electrophilic reaction at three of its ring C atoms, **5**. The accumulation of negative charge is the target for the incoming nitronium ion. Note that the electron cloud accumulates at the positions that will result in the symmetrical TNT molecule you are aiming to make.

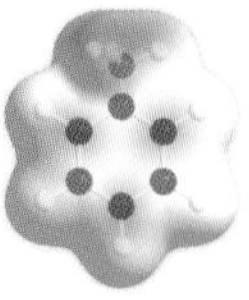

Now we get into position to watch the reaction (Figure 16.2). As we watch, the nitronium ion sniffs around and homes in on one of the electron-bloated C atoms of the ring. The latter's partial negative charge acts as a beacon to guide in the positively charged attacker. After bustling through the mixture the ion arrives at the molecule. We see the electrons of the ring respond to the presence of the positive charge on the N atom when it gets very close. Two electrons that originally contribute to the carbon–carbon double bond at that point bulge out of the ring, like the projection of an amoeba, and form a carbon–nitrogen bond. This bond formation disrupts the distribution of electrons around the ring because the alternation of single and double bonds is now incomplete.

To recover this energetically low arrangement of bonds something has to give. In this case, we see the H atom at the site of substitution release its pair of electrons and it escapes into the solution as a proton. That pair rejoins the ring and the alternation of single and double bonds is re-formed. The result is singly nitrated toluene.

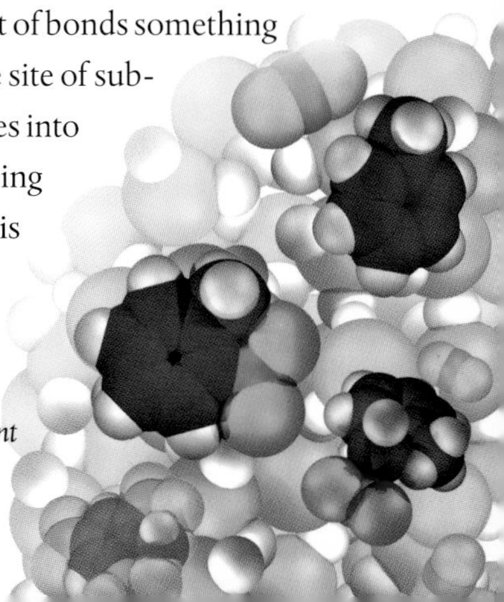

FIG. 16.2 *The nitration event*

If we stay and watch, we see a similar sequence of events, with the nitronium missile attacking the two other electron-rich regions of the molecule. In practice, the industrial production of TNT has to take precautions to remove some of the by-products, such as those formed by nitration at the 'wrong' C atoms which, as I mentioned, can lead to an unstable compound.

Molecular convulsions

Perhaps we can relax for a moment and use the time to see why TNT is such an effective explosive. One advantage of TNT is that it is reasonably stable against shock and can be handled safely, even when molten. However, let's stand as close as we dare to a TNT molecule and observe what happens when the solid is detonated (Figure 16.3). Detonation is simply a short, sharp molecular shock caused, for instance, by the electrically induced explosion of a less stable compound nearby. We see the TNT molecule respond by convulsing, twisting and turning, flexing violently, and literally shaking itself apart into lots of fragments as the atoms link together as little molecules such as N_2, CO, and H_2O. Even individual C atoms fly off and cluster together as soot. We are caught in a devastating maelstrom as the compact solid is suddenly replaced by a highly compressed gas of small gaseous molecules. The sudden formation of gas and its immediate expansion generates the destructive shockwave of the explosion.

The generation of C atoms, congregated as soot, is a characteristic of a TNT explosion. The soot can be eliminated, more gas generated, and the explosion thereby rendered more violent, if an oxidizing agent is added to the mixture. In that case, more of the carbon is oxidized to CO_2. The addition of the oxidizing agent ammonium nitrate to TNT gives the military explosive amatol.

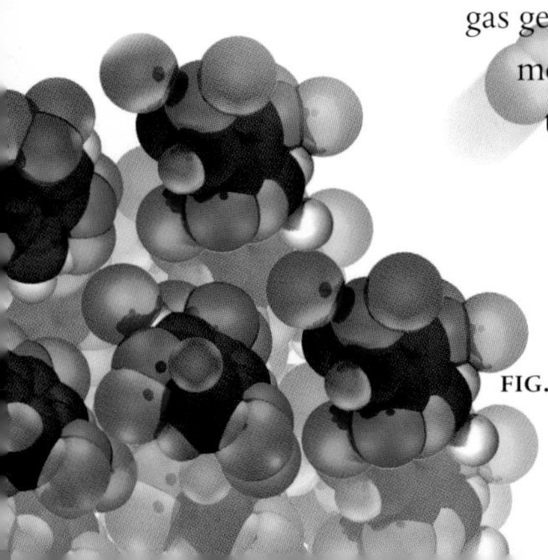

FIG. 16.3 *Explosion of TNT*

Variation on a theme

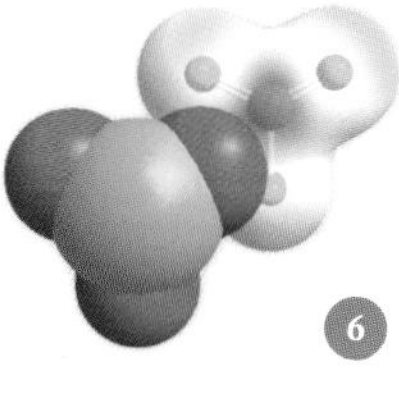

So much for the attachment of an N atom to a molecule. Suppose you want to decorate your benzene building with sulfur, S, perhaps for the synthesis of a drug. You should suspect that that

6

can be achieved by using an electrophilic missile, but one with sulfur in its warhead rather than nitrogen. One approach is to use sulfuric acid alone rather than a mixture with nitric acid. The reaction is faster, though, if instead of sulfuric acid itself you use 'oleum', which is a vicious (and viscous) liquid in which sulfur trioxide, SO_3, has been dissolved in the concentrated acid. The SO_3 molecule, **6**, is the electrophile. It acts in that way on account of the ability of the O atoms to suck electrons away from the S atom, so leaving the S atom's nuclear charge to shine through the wispy remaining electron cloud. In other words, the S atom has a partial positive charge and can use it to sniff out the partial negative charge of the dense regions of electron clouds on other molecules. Oleum can be used to attack even undecorated benzene, so let's consider that reaction.

When we watch the reaction, we see events very similar to those accompanying the formation of TNT (Figure 16.4). The positively charged S atom of the SO_3 molecule is attracted to the reasonably dense electron cloud on a C atom of the ring. When it arrives we see it suck out a pair of electrons that initially contributed to the carbon–carbon double bond. At this point the molecule has a carbon–sulfur bond. However, there is still an H atom attached to the same atom, so the stabilizing alternation of double and single bonds in the ring has been lost. As we watch we see a nearby sulfuric acid molecule respond to the partial negative charge on the O atoms of the attached $-SO_3$ group and flip across a proton. The whole ring now shudders as the

FIG. 16.4 *Sulfur attaching*

electron cloud readjusts to the shock of its arrival and the charge it brings. We see the H atom that shares the C atom being expelled and the alternation of single and double bonds restored.

Wolves dressed as sheep

The nature of the starting material, the initial substituted benzene, plays a vital role in the reaction. You should be able to anticipate that if a group of atoms already present pumps electron cloud into the ring, then the rate of electrophilic attack will be increased. The opposite is true if the group sucks electron cloud out of the ring. Moreover, that pumping and sucking might be selective and affect particular positions around the ring in different ways. If you compare the electron clouds of phenol, 7, and benzene, 8, it is apparent that alternating positions in the former have a higher density of electron cloud on the C atoms than in benzene, so you should expect it to undergo electrophilic attack more readily.

This gives me the opportunity to introduce an electrophile with that character subtly concealed. A molecule of bromine, Br_2, is an electrophile, although without any obvious charge, it looks more sheep than wolf. However, wolf under sheep it is, on account of the ability of its numerous electrons to slurp around and transitorily expose the nucleus of each atom. Think of a swirling fog, with first one bromine nucleus dimly exposed and the other more heavily concealed, and then another swirl that reverses the first. The first of these transient arrangements can be induced to form more strongly and then survive if one of the atoms is close to an electron-rich region of the target molecule. You saw in Reaction 15 how an incoming missile could induce its target to form; here the innocent target is preparing the incoming missile for its own demise.

REACTIONS

Bromine does not attack benzene itself. However, it reacts readily with phenol due to the accelerating effect of the –OH group attached to the ring. The acceleration is even more pronounced if the –OH group of phenol is replaced by an –NH$_2$ group, to give the compound aniline, with an even greater accumulation of electron density at the alternating positions on the ring, 9.

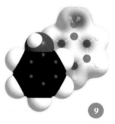

9

Masking tape

As a determined but cautious molecular architect you might wonder if the acceleration can be brought under control so that substitution does not occur at all the relevant locations but at only one. Suppose, for instance, you wanted to attach an atom to the C atom diametrically across the ring from the –NH$_2$ group in aniline and not to the two neighbouring C atoms. One approach is to attach an electron-cloud-sucking group to the electron-cloud-pumping –NH$_2$ group. Then fewer electrons will be available to be transferred to the benzene ring. A secondary effect of the newly attached group might be to shelter the C atoms next to the one to which the N atom is attached, so attack is less likely to occur there (Figure 16.5). You can think of the procedure as akin to finding a way to cover up part of the molecule with masking tape so that when you spray it with electrophile only the unexposed part is decorated. Once again, you are beginning to see how chemists, although working in flasks and beakers, are able to guide atoms to particular destinations and steer them away from others.

Suppose you want to attach C atoms to build up the carbon network. How can that be done? I introduce that procedure in Reactions 21 and 22.

FIG. 16.5 *Protection*

17

Fasteners

ACID CATALYSIS

I explained the general basis of catalysis in Reaction 11, where I showed that it accelerated a reaction by opening a new, faster route from reactants to products. One of the ways to achieve catalysis in organic chemistry is to carry out a reaction in an acidic or basic (alkaline) environment, and that is what I explore here. In Reaction 27 you will see the enormous importance of processes like this, not just for keeping organic chemists productive but also for keeping us all alive; I give a first glimpse of that later in this section too. Various kinds of acid and base catalysis, sometimes both simultaneously, are going on throughout the cells of our body and ensuring that all the processes of life are maintained; in fact they are the very processes of life. I deal with acid catalysis in this section and base catalysis in the next.

The point to remember throughout this section is that an acid is a proton donor (Reaction 2) and a proton is an aggressive, nutty little centre of positive charge. If a proton gets itself attached to a molecule, it can draw electrons towards itself and so expose the nuclei that they formerly surrounded. That is, a proton can cause the appearance of positive charge elsewhere in the molecule where the nuclei shine through the depleted fog of electrons. Because positive charge is attracted to negative charge, one outcome is that a molecule may be converted into a powerful electron-sniffing electrophile (Reaction 16). Another way of looking at the outcome of adding a proton is to note that a C atom with a positive charge is a target for nucleophilic missile attack (Reaction 15). Therefore, if a proton draws the electron cloud away from a nearby atom, then its presence is like a fifth-column agent preparing a target for later attack.

Joining up

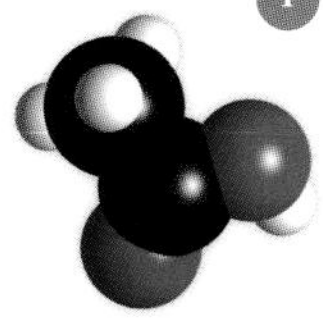

Let's shrink and watch as some acid is added to a molecule that contains a –CO– group, such as acetic acid (I, Figure 17.1). The protons provided by the added acid are riding on water molecules, as H_3O^+ ions, and arrive in the vicinity of the acetic acid molecule. Once within hurling distance, an ion transfers its spare proton to the –CO– group. There is a flurry of activity as the electron cloud of the acetic acid molecule responds to the arrival of the proton, and when it settles down (in about a thousand-trillionth of a second) we see a glint of the positive charge of the O atom shining through the depleted cloud at that atom. The acetic acid molecule has been converted into an electrophile. The positive charge on the O atom can sniff out, dog-like, any regions of high electron density on other molecules and go about its chemical business.

As I remarked, as well as being converted into an electrophile, under the influence of a proton a molecule can also become the target

FIG. 17.1 *Proton transfer to acetic acid*

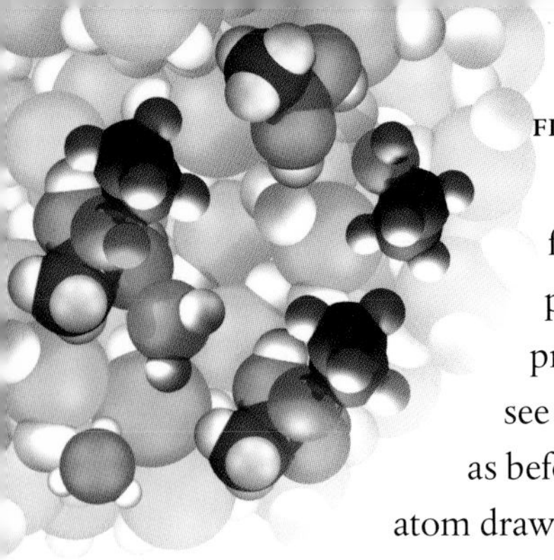

FIG. 17.2 *Reaction with ethanol*

for nucleophilic attack. Let's watch what happens when ethanol (common alcohol, 2) is also present in the mixture around us (Figure 17.2). We see a proton attach to the acetic acid molecule just as before. The positive charge that it creates on the O atom draws some electron cloud away from the neighbouring C atom and the latter's nuclear charge shines through and acts as a beacon to attract a nucleophile. We see an alcohol molecule wander into the vicinity, hesitate, and then move towards the positively charged C atom.

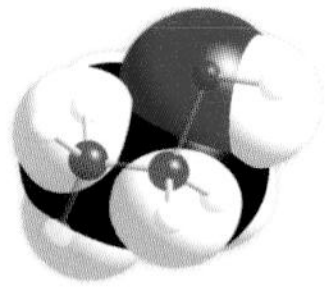

2

An alcohol molecule is a nucleophile by virtue of the dense electron cloud on its O atom, which results in a partial negative charge there. As soon as it gets within striking distance of the target molecule the electron cloud bulges out towards the positively charged C atom and forms a carbon–oxygen bond. At about the same time, the alcohol molecule shrugs off the H atom (as a proton) that it brought with it; that proton is picked up by a nearby water molecule, which becomes an H_3O^+ ion. A proton has been restored to the medium: one was used to prepare the target and now one has been returned. That marriage-broker-like action of the proton, its role in bringing about reaction but its release for involvement elsewhere after its work has been done, is the central nature of catalytic action (Reaction 11).

The molecule that we are watching being formed doesn't survive for long, for the proton that helped in its creation now changes its spots and aids in its destruction. We watch it happen (Figure 17.3). First, we see an H_3O^+ ion (not necessarily the one so recently formed) donate a proton to one of the −OH groups of the newly formed molecule. We see that the newly protonated part of the molecule strongly resembles an H_2O molecule but one with a positive charge. As we watch, we see the electron cloud

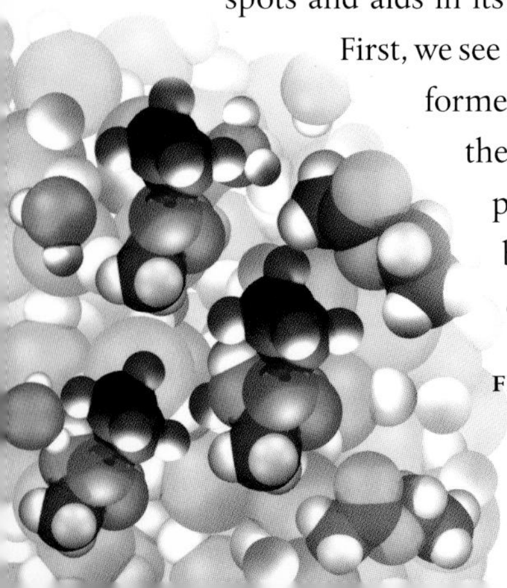

FIG. 17.3 *Making an ester*

of the entire molecule shudder, the shared pair of electrons holding it to the rest of the molecule snap off the C atom in response to the charge, quench it, and the water-like H_2O group shake off and wander off as actual water amid the myriad other molecules of the mixture. That step has stripped the molecule of one of its O atoms.

There is more to come. The rearrangement of electrons has left the oxygen nucleus of the remaining –OH group exposed and that atom has a positive charge. It can get rid of it by shuffling off the H atom as a proton. We see this happen, and a proton is restored to the medium (as an H_3O^+ ion) to make up for the one that was used to initiate the destruction. So, we have watched another marriage-broking, or more specifically divorce-broking, catalytic step.

I have referred to the second step as destructive, but in fact a new substance, 3, has been built from the reagents, and that substance is highly important. It is an 'ester', a compound formed from the reaction of a carboxylic acid (an acid with a –COOH group, like acetic acid) and an alcohol (a compound with just an –OH group, like ethanol). Esters contribute considerably to the world around us. For instance, many fruit flavours are due to esters, fats and edible oils are esters, many foodstuffs have components composed of ester-like molecules (I introduce one below), and many of the processes going on inside your body involve the formation and destruction of esters.

Pedant's point. The origin of the name 'ester' is a contraction of the German *Essigäther*, from *Essig* 'vinegar' (acetic acid) and *Äther* 'ether'. Whoever coined the term in the late nineteenth century didn't quite understand the nature of the compound he—and then it would have been a he—was naming.

Breaking apart

Acids can also catalyse the decomposition of an ester into its parent carboxylic acid and alcohol. That is, they can bring about the reverse of the construction reaction I have just described. They catalyze a 'hydrolysis reaction', the severing of a bond by introducing the elements of water (that is, two H atoms and one O atom, as in H_2O). All you have to do is to imagine the steps I have described as running in reverse. As I shall illustrate in a moment, steps like the ones we are about to observe take place when you eat.

In the interests of reality and relevance, let's drop into your own stomach, immerse ourselves in the slop there, shrink to molecular size, and watch what happens when you eat apple pie. I shall focus on two components of the masticated chunks that are splashing down on us. First, I shall focus on one of the apple flavour molecules, an ester with a reasonably long chain of carbon and hydrogen atoms; then I shall consider your digestion of the cane sugar used in the preparation of the pie crust. A lot of other stuff is going on, of course, but the gastro-history of these two components will give you an idea of what is happening, even as you read these words.

The liquid in your stomach is acidic, which means that H_3O^+ ions are abundant there. That acidity could be a problem for your continued existence, but the walls of your stomach are cleverly designed, that is, evolved, to avoid it digesting itself.

It doesn't take long before we catch sight of an H_3O^+ ion tossing its spare proton across onto a nearby ester molecule (Figure 17.4). As I have already explained, that proton's charge drags electrons from the C atom that throughout our watching will be the centre of the action, and thereby makes it a target for nucleophilic attack. The obliging nucleophile is nothing other than one of the ubiquitous water molecules that are present in the sloppy,

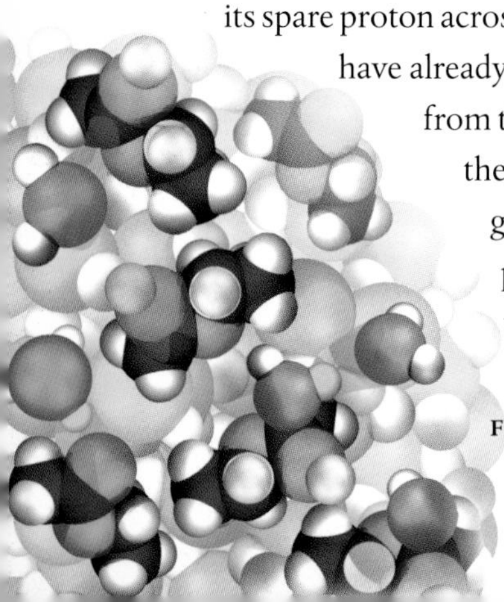

FIG. 17.4 *Breaking up an ester*

FIG. 17.5 *Final stage of breaking an ester*

complex medium. The dense electron cloud on the O atom of a water molecule sniffs out the positive charge on the ester's C atom, an amoeba-like process of electron cloud bulges out from its O atom, and forms a bond. We see one of the protons of the newly arrived H_2O molecule skip away onto a neighbouring water molecule, so that little catalytic cycle is over, with an −OH group now attached to the C atom.

Now the second cycle begins as you continue your digestion by using another proton (Figure 17.5). This time a proton from a nearby H_3O^+ ion lands on the O atom that is not part of an −OH group. The electronic convulsion its arrival causes results in the severance of the carbon–oxygen bond and the springing away of an alcohol molecule. That molecule is free to go on its path to further digestion. Then we see a final little shudder running through the electron cloud as the

remaining molecule sheds another proton and becomes a carboxylic acid that wanders off for further metabolism. The release of the second proton completes the second catalytic cycle. The outcome is a carboxylic acid and an alcohol.

Another component of the chink of masticated apple pie that dropped into your stomach is a sugar molecule. A sucrose (cane sugar) molecule is a combination of a glucose molecule and a fructose molecule, 4. Hydrolysis snips the two components apart and makes these much smaller molecules available for powering our muscles and thoughts. An alternative fate is their conversion to fat-like molecules and their storage as contributions to obesity.

As we watch we see a proton attach to the O atom that acts as a bridge between the two components of this big lumbering molecule (Figure 17.6). Its positive charge sucks electrons away from the

FIG. 17.6 *Cutting up a carbohydrate*

adjacent C atom of the fructose molecule, rendering it susceptible to attack by an H_2O molecule acting as a nucleophile. We see the molecule jostle in and an amoeba-like prong of electron cloud gropes out towards the C atom to form a carbon–oxygen bond. The formation of that bond drives out the O atom belonging to the fructose ring, which is left with a positively charged H_2O group. But the molecule makes short shrift of that, and we see it transfer a proton to a nearby water molecule, which leaves the fructose molecule with an −OH group in the right place. The end result of this step in the highly complex series of steps that we call digestion and which fuels our growth and actions is the scission of a sucrose molecule into its glucose and fructose components, which go on their way to further metabolism.

Casting pearls

Now let's see what happens in the dark, dank, slightly bubbly interior of your stomach when you eat meat or, if that doesn't appeal, then protein of some kind. As I remarked in Reaction 14, proteins are elaborate forms of nylon. You could think of a protein as a bundled string of pearls, each pearl being a special group of atoms and the links

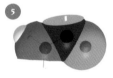

between them are 'peptide links', −CONH− , 5. I shall deal with real proteins later (Reaction 27); to keep things simple at this stage, I shall describe how the peptide link is broken down by acid hydrolysis, but will use a very parsimonious string of just two pearls, 6.

Pedant's point. The pearls of actual proteins are amino acids. The two-pearl protein model I am discussing is more properly called an 'amide'.

One problem with our model, which is also an aspect of all proteins, is that the peptide link is sturdy and hard to break apart. For proteins, and the stability of our flesh, our hair, our beaks, and our claws, that is an advantage. That sturdiness is one reason why when we cook meat we don't end up with a gruesome slurry of foul smelling and poisonous molecules. As a result, rather severe conditions are needed to break the link, such as boiling with sulfuric acid for several hours.

The origin of this stability will guide our thoughts about the course of the reaction, so I need to spend a moment on it. When the link is examined you see that the electron cloud is surprisingly dense in the region between the N atom and the C atom of the –CO– group, 7. That accumulation of electron glue makes the bond strong and resistant to attack. The extra electrons (which make it something akin to a double bond) have been drawn in from the N atom, so that the atom has a slightly thinned electron cloud. By now, you know the consequence of thinned electron clouds: the positive charge of the nucleus glints through, and the N atom is slightly positively charged. The migration of the electrons has a second consequence: it pushes the electrons in the region of the carbon–oxygen bond on to the O atom itself. As a result of the accumulation of electron cloud, the O atom becomes slightly negatively charged. That is a crucial point for what follows.

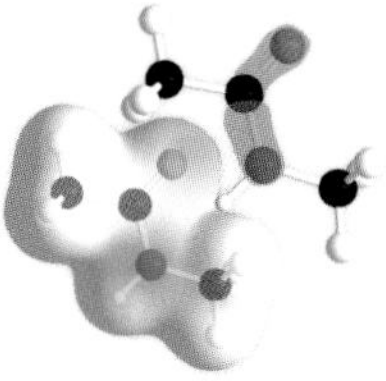

We are back as molecular onlookers in your acidic and turbulent stomach again (Figure 17.7). Just as in the case of ester hydrolysis, there are two cycles of acid catalysis when you start to digest a protein, the snipping apart of two pearls. Because, as I have just explained, the O atom of the link is slightly negatively charged, you can now appreciate that in an acid medium it will be the O atom that accepts a proton. Once it has done

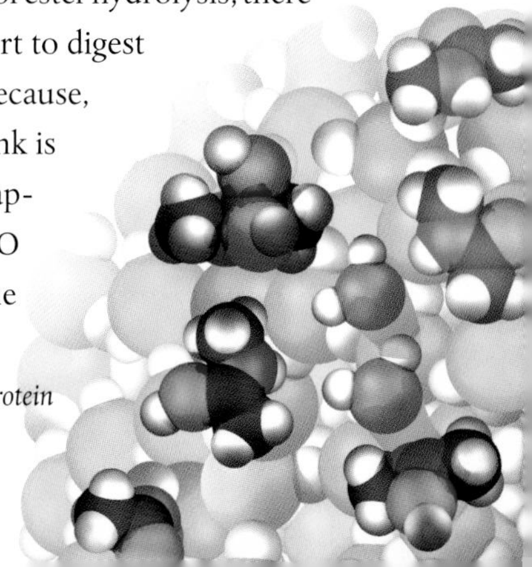

FIG. 17.7 *Cutting up a protein*

so, electron cloud is pulled away from the C atom attached to the CO group and it becomes partially positively charged. This charge makes the atom a target for an incoming water molecule acting, through the lone pair on its O atom, as a nucleophile. We see the electron cloud on the O atom of the incoming H_2O molecule bulge out towards the C atom and a carbon–oxygen bond form. The additional proton skips away onto a neighbouring water molecule.

At this point we see a proton jump onto the N atom. The electron cloud shudders under the impact of its arrival, the carbon–nitrogen bonds breaks, and the two parts of the molecule, the two pearls of the tiny chain, drift apart. The protein is on its way to digestion, with its precious components perhaps destined to be linked into another string of pearls to help you think, grow, or reflect on the private life of the atoms within you.

Zippers

BASE CATALYSIS

A base, you should recall from Reaction 2, is the second hand clapping to the acid's first. That is, whereas an acid is a proton donor, a base is its beneficiary as a proton acceptor. The paradigm base is a hydroxide ion, OH⁻, which can accept a proton and thereby become H_2O. However, in the context of catalysis, the topic of this section, its role is rather different: instead of using its electrons to accept the proton, it uses them to behave as a nucleophile (Reaction 15), a searcher out of positive charge. Instead of forming a hydrogen–oxygen bond with an incoming proton, it sets the electronic fox among the electronic geese of a molecule by forming a new carbon–oxygen bond and thereby loosening the bonds to neighbouring atoms so that they can undergo rearrangement. The OH⁻ ion in effect unzips the molecule and renders it open to further attack.

Base catalysis has a lot of important applications. An ancient one is the production of soap from animal fat. To set that scene, I shall consider a simple model system, the 'hydrolysis' (severing apart by water) of the two components of an ester, **1** (the same compound I used in Reaction 17, a combination of acetic acid and ethanol), and then turn to soap-making itself. You saw in Reaction 17 how esters can be broken down into their components, a carboxylic acid and an alcohol, by an acid; here we see the analogous reaction in the presence of a base. To be specific, the reagent is a solution of sodium hydroxide, which provides the OH⁻ ions that catalyse the reaction.

Unzipping esters

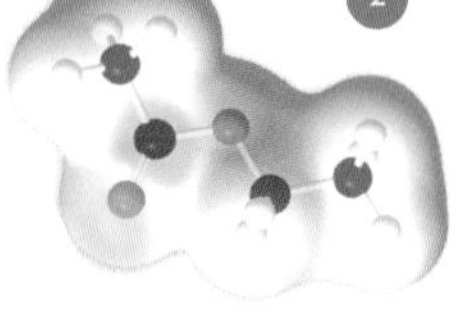

We watch what happens when a solution of sodium hydroxide is added to an ester and the mixture is boiled (Figure 18.1). The O oxygen atoms of the ester have already ripened the molecule for nucleophilic attack by drawing some of the electron cloud away from the C atom to which they are both attached, leaving it with a partial positive charge, **2**. The negatively charged OH⁻ ion sniffs out that positive charge and jostles in to do its business. Once within bonding distance, its electron cloud bulges out towards the C atom and forms a carbon–oxygen bond. The electrons in the vicinity of that C atom then shake themselves down into an energetically more favourable arrangement.

In that shaking down, as the electron cloud presses in from the incoming OH⁻ ion, the cloud that forms the double carbon–oxygen bond of the ester retreats onto the O atom, but then surges forward again like an incoming tide to reform the original strong double bond. To form that bond, the C atom must relax its hold on the electrons in the bond that held the alcohol frag-

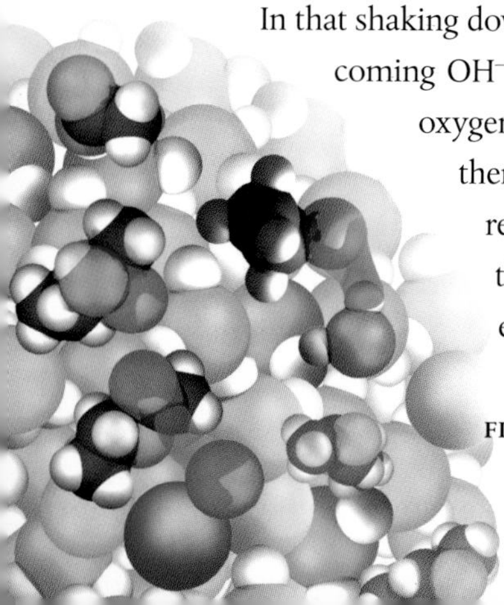

FIG. 18.1 *Breaking up an ester*

ment in place, and that bond breaks. The electron cloud responsible for that bond retreats onto the alcohol's O atom and gives it a negative charge. The liberated molecule drifts away, but almost immediately extracts a proton from a nearby H_2O molecule, converting that molecule into an OH^- ion. The cata-lytic cycle has been completed: the OH^- ion used for the initial attack has been restored. The molecule left after the alcohol has departed in acetic acid, the acid originally formed to make the ester. The outcome, though, is separate acid and alcohol molecules. The bond between them has been severed: the ester has been hydrolysed.

Now let's consider soap itself. Tallow, beef fat, is an ester, 3, in which a glycerol molecule, 4, is combined with three long-chain car-boxylic acids, stearic acid ('stear' is the Greek word for tallow). When boiled with sodium hydroxide, it undergoes base-catalysed hydroly-sis in exactly the same way as we have just watched for the model ester. The attack by the OH^- ions in the solu-tion severs the carboxylic acid chains from the glycerol, leaving glycerol itself and, more importantly, long-chain carboxylic acid molecules (as carboxylate ions, ending in a $-CO_2^-$ group in place of the $-COOH$ group of the acid).

A soap molecule acts by virtue of its Janus-like character: the long chain hydrocarbon tails can sink into blobs of hydrocarbon grease and the 'head groups', the $-CO_2^-$ groups of the carboxylate ions, remain on the surface of the blob. These head groups become sur-rounded by water, so the blob can be washed away.

Unzipping proteins

You will recall the two-pearl string of molecules that I used as a simple model of a protein in Reaction 17 (I reproduce it here as 5). There you saw that the

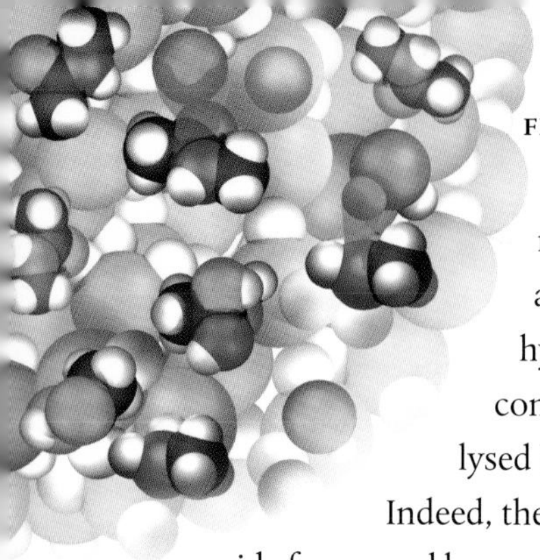

FIG. 18.2 *Breaking up a protein*

metaphorical pearls could be cast apart by the action of an acid that acted as a catalyst for the hydrolysis of the bond that joins them. It should come as no surprise that the link can also be hydrolysed by using a base as a catalyst in place of an acid. Indeed, they are attacked by boiling with sodium hydroxide for several hours, and the molecular mechanism is very similar to that involved in the base hydrolysis of an ester.

We can watch, to make sure (Figure 18.2). As for the ester, we see a hydroxide ion, OH⁻, sniff out the region of depleted electron density on the C atom of the –CONH– link and form a bond to it. The electron cloud rearranges in a copycat version of the ester hydrolysis, and the two pearls drift apart. One of them has a negative charge on the N atom, but that is soon quenched by the arrival of a proton from a nearby H_2O molecule, which is thereby converted into an OH⁻ ion. The catalytic cycle is complete: an OH⁻ ion has been used but restored. We are left with the separated pearls.

A question that might be forming in your mind is, given that severe conditions—boiling in alkali for hours in this instance and boiling in acid for hours in Reaction 17—are needed to break up the components of the model compound and therefore presumably a real protein too, how is it that you can digest the proteins that you ingest as food? After all, digestion begins in the relative damp, cool environment of your mouth and continues in the damp, cool, and mildly acidic conditions of your stomach. Nature had to overcome a major problem: she had to achieve proteins that are stable enough to constitute one organism for a lifetime but unstable enough for another organism to eat them. As usual, she stumbled onto a remarkable answer, which I explain in Reaction 27.

Adding Up
ADDITION

A
dding up, or more formally 'addition', is just what it says: it is the attachment of atoms to a sensitive spot on a molecule. I need to stand back for a moment and explain what I mean by a 'sensitive spot'. You already know (from Reaction 13) that some atoms are held to each other by an electron cloud (a 'single bond') and others by a doubly dense cloud (a 'double bond'). There is a third type in which the atoms are held together by an even denser, triply dense cloud of electrons, forming a 'triple bond'. Here I am concerned with the latter two types of bond, the so-called 'multiple bonds'. These are the sensitive spots of organic molecules for it is quite easy to attack a multiple bond, rearrange the clouds, and attach other groups. For simplicity, I shall deal only with the more common type of multiply bonded molecule, one with a double bond.

A double bond is a region rich in electrons, so you should suspect that any missile that will attack it will be an electrophile (a seeker-out of negative charge, of electron richness, Reaction 16). I shall consider a very simple case: the addition of bromine to cyclohexene, **1**. As we have seen in Reaction 15, the presence of a bromine atom, Br, in a molecule is often the starting point for building on other groups of atoms, so this is an important reaction in a chain that might be used to construct something useful, such as a pharmaceutical. Bromine is a liquid composed of Br_2 molecules. Cyclohexene is a liquid composed of hexagonal benzene-like molecules but with only one double bond in each molecule. Why I have chosen this slightly elaborate molecule rather than something simpler, such as ethylene (ethene, Reaction 13), will soon become clear.

Let's shrink together down to our normal molecular size and watch what happens as the bromine is poured into the cyclohexene (Figure 19.1). We already know from Reaction 16 that a bromine molecule has a nose for negative charge, so you should not be surprised to see one homing in on the electron-rich double bond of a nearby cyclohexene molecule. We see a lot of rearrangement of the electron cloud once the bromine molecule is in contact with the double bond. In particular, some of that bond's electron cloud bulges out from one C atom towards the exposed nucleus of the incoming molecule and starts to form a bond to it instead of to its original partner C atom. At the same time as that carbon–bromine bond starts to form, we see the original bromine–bromine bond starting to lengthen and weaken. At the end of this skirmish, a new carbon–bromine bond has formed, the old carbon–carbon double bond has dwindled to a single bond, the original bromine–bromine bond has broken, and the spare Br atom has drifted away as a Br⁻ ion. Gone it might be, but it has duty to do shortly.

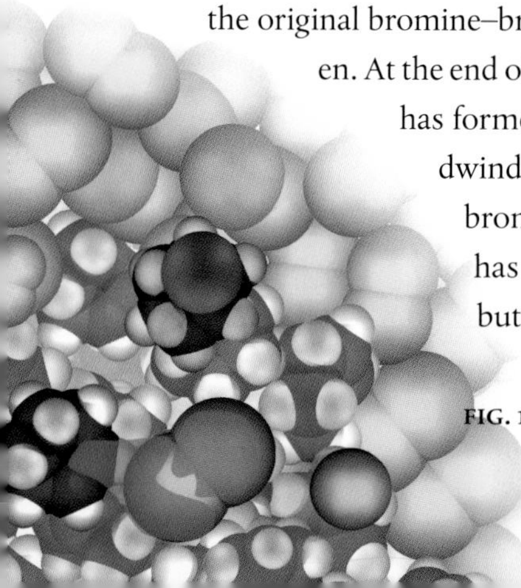

FIG. 19.1 *Starting addition*

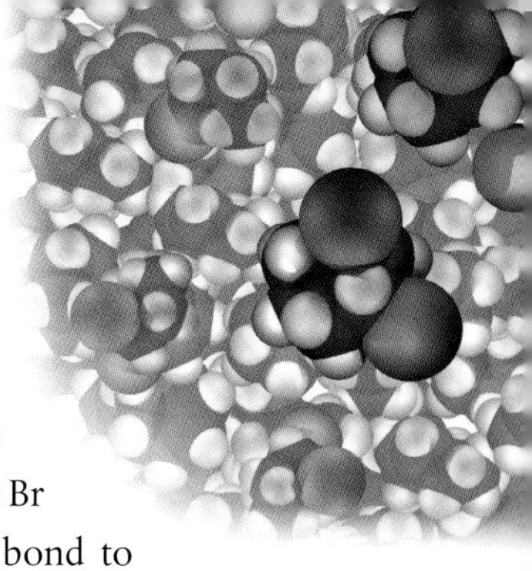

FIG. 19.2 *Completing addition*

The second C atom of the original double bond has lost its share in the electron cloud and has thereby acquired a positive charge. The nearby newcomer Br atom is electron-rich. We see the inevitable outcome: the electron cloud on the Br atoms snaps across to form a carbon–bromine bond to that C atom without breaking the initial carbon–bromine bond. As a result, the C, Br, and C atoms form a little local triangle.

The Br⁻ released in the original attack is swimming around in the vicinity and others from similar reactions nearby are too. We see one of these ions being attracted to the positive charge of our cluster of atoms (Figure 19.2). It swoops in, drawn in by the attraction between opposite charges, and locks on to the second C atom. The first carbon–bromine bond of the triangle springs open, and the product forms.

There is an important point. The Br⁻ ion cannot bond to the molecule if it approaches from the side to which the first Br atom is attached because that atom shields the C atoms. The only side from which it can approach is the side opposite to the resident Br atom. Therefore, instead of getting a mixture of products with the two Br atoms attached on either face of the molecule, there is only one with the atoms attached on opposite sides. (This is the reason why I chose cyclohexene: it can't twist into a different shape, so the distinction between the products is preserved.) Once again, you see that it is possible to achieve subtle control of molecular architecture. Put another way: Nature forbids the achievement of what she is disinclined to allow.

Taking Away

ELIMINATION

The hot spots of molecules that I have identified as double bonds (two shared pairs of electrons lying between the same two carbon atoms) and their triple bond cousins are often desirable entities. They are desirable either in their own right or because they can be used in the course of the construction of an elaborate molecule. For instance, a double bond can make the molecule stiffer and resistant to twisting. In Reaction 28 you will see that one particular natural product, quinine, must have a double bond in a particular position for it to be able to function—Nature is very particular about the shape of a molecule that she uses—and the drug's synthesizers had to find a way to introduce it. How, though, can a double bond be introduced into a molecule that begins life with only single bonds? One approach is 'elimination', the expulsion of groups of atoms on neighbouring carbon atoms, leaving those two atoms free to form a second or even third bond to each other.

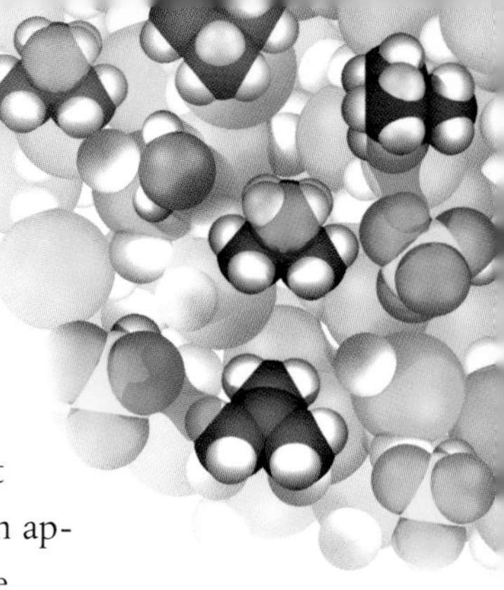

FIG. 20.1 *Creation of a double bond*

One approach is to pull an H atom (as a proton) or some other group of atoms off one C atom, and then hope that the ensuing convulsions of the electron cloud will result in its accumulation to form a double bond between that C atom and its neighbour. There are two common approaches, one involving an acid and the other a base.

Let's watch what happens when sulfuric acid, **1**, is added to **2** (Figure 20.1). The acid, a proton donor, generates H_3O^+ ions in the usual way by transferring a proton to a neighbouring water molecule and leaving behind an HSO_4^- ion, and we see one of these ions sidle up to the target molecule. A proton hops across onto the O atom from the H_3O^+ ion, so forming a positively charged $-OH_2^+$ group. There is an immediate convulsion of the electron cloud, and that group escapes as an H_2O molecule, leaving behind a positively charged ion with the positive charge mostly localised on the C atom. This ion is un-

stable but survives briefly. Before long we see a proton hop off a C atom on to a nearby HSO_4^- ion left behind when a sulfuric acid molecule donated a proton. The part of the molecule's electron cloud that held that proton withdraws towards the positively charged C atom. That cloud now lies between two C atoms and corresponds to the formation of a double bond.

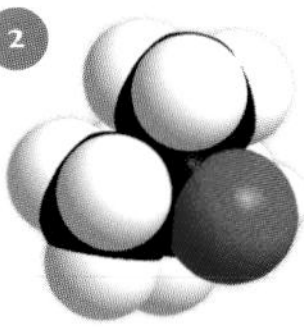

Now consider a similar starting molecule, **3**, but with a Br atom in place of the OH group. In this case, you use a base to carry out the reaction and something slightly different happens.

We watch as an OH^- ion, a seriously effective proton acceptor, drifts up to the molecule (Figure 20.2). Once it is in contact, we see it begin to extract a proton from a $-CH_3$ group.

FIG. 20.2 *Another way to create one*

This is no easy task, but it is assisted by the obliging electron-hungry Br atom. We see the carbon–bromine bond lengthening as the Br atom acquires a full share of the electron cloud and begins to slip quietly away as a Br^- ion. The electron cloud that initially glued the proton to the C atom become free to seep towards the other C atom and start to form a double bond. What we have seen is the gradual removal of a proton aided by the departure of a complaisant Br atom, and the outcome is a carbon–carbon double bond.

You might have identified a problem: in each case the reaction is on the brink of becoming a substitution of the kind I described in Reaction 15. One slip, and 2 could become not the elimination product but the substitution product with a sulfur-containing bundle of atoms attached to it. Similarly, one slip, and 3 could become not the elimination product but a substitution product with an OH group attached. There is a fine balance between an ion—in these cases HSO_4^- and OH^-—acting as a base and extracting a proton and instead acting as a nucleophile and bringing about substitution. Put another way, there is a fine balance between ions eyeing up and attacking a proton, resulting in elimination, and eyeing up and attacking a C atom, resulting in substitution. Chemists have to judge carefully how to favour one type of outcome over another and select the conditions appropriately. The HSO_4^- ion from sulfuric acid, with its negative charge spread thinly over all four O atoms, for instance, is a very poor nucleophile, so it is unlikely to result in substitution.

Carbon Footprints

THE WITTIG REACTION

As a molecular architect working on an atomic construction site you need to be able to build up the carbon skeleton of your project, not merely decorate it with foreign atoms. There are dozens of different ways of doing that, and in this and the next section I shall introduce you to just two of them to give you a taste of what is available.

A secondary point is that throughout chemistry you will find reactions denoted by proper nouns, recognizing the chemists who have invented or developed them. One example is that of the 'Wittig reaction', which is named after the German chemist Georg Wittig (1897–1987; Nobel Prize 1979). The reaction is used to replace the oxygen atom of a CO group in a molecule by a carbon atom, so that what starts out as decoration becomes part of a growing network of carbon atoms.

You need to know that phosphine, PH_3, **1**, the phosphorus cousin of ammonia, NH_3, is a base (Reaction 2). When it accepts a proton it becomes the ion PH_4^+. The H atoms in that ion can be replaced with other groups of atoms. A replacement that will be of interest is when three of the H atoms have been replaced by benzene rings and the remaining H atom has been replaced by $-CH_3$. The resulting ion is **2.** In the presence of a base, such as the hydroxide ion, OH^-, the $-CH_3$ group can be induced to release one of its protons, so the positive ion becomes the neutral molecule, **3.** Note that there is a partial positive charge on the P atom and a partial negative charge on the C atom of the CH_2 group. The presence of that partial negative charge suggests that the species could act as a nucleophile (Reaction 15), a seeker out of positive charge, with the CH_2 group the charge-seeking head of the missile.

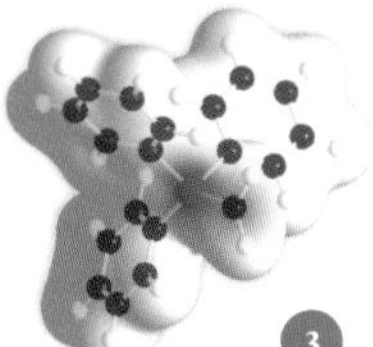

Let's watch what happens when **3** attacks a molecule with a CO group, specifically **4**: perhaps you want to sprout a carbon chain out from the ring and intend to begin by replacing the O atom with a C atom. If you examine the molecule closely, you will see that the O atom has sucked some electron cloud away from the neighbouring C atom. As a result, the positive charge of the carbon nucleus shines through the depleted electron cloud and acts as a homing beacon for the negative charge on the CH_2 group of the incoming phosphorus missile. As we watch (Figure 21.1), we see the electron cloud bulge out from the CH_2 group, reach towards the carbon nucleus, and form a carbon–carbon bond.

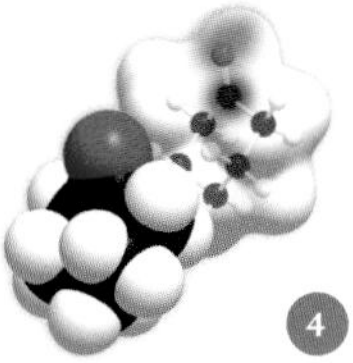

You have achieved what you wanted: a slightly longer carbon side chain. But there is a great deal of baggage attached to it. You need to get rid of it and shave

FIG. 21.1 *Joining two molecules*

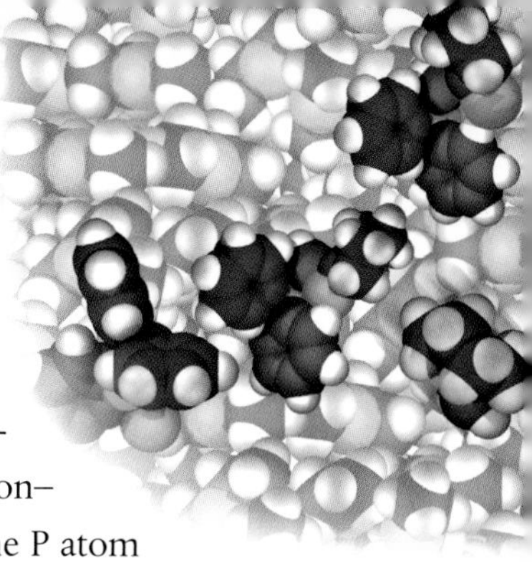

FIG. 21.2 *Shaking off the baggage*

the side chain back to just the C atoms. The cleverness of Wittig's procedure is that the molecule shaves itself.

Let's go on watching the newly formed molecule (Figure 21.2). We see that when the new carbon–carbon bond is formed, the positive charge of the P atom is brought close to the negative charge that is now on the O atom. We see them snap together to form a phosphorus–oxygen bond, so completing a four-membered ring. For reasons related to the way that electron clouds can arrange themselves around atoms, six-membered rings, like that of benzene, are reasonably stable, but four-membered rings, like the one we have just seen formed, strain to exist because the electron cloud can't adjust into the optimal position for forming bonds. However, the stressed ring can relax by reorganizing its electrons.

We see something extraordinary happening. The electrons of the ring convulse and shake off the entire P atom and its baggage, which includes the O atom of the original molecule. This is a common theme in organic chemistry: the phosphorus–oxygen bond is so strong that P atoms can be used to pluck O atoms out of molecules rather like strong bases can be used to pluck out protons.

Once you have seen what is going on, you should be able to see how to build more complicated molecules. As you have seen, the C atom that has replaced the O atom is supplied by the phosphorus compound. Therefore, if you start with a phosphorus compound that has a more elaborate group in place of the CH_2 group, then that more elaborate group will replace the O atom. Your construction business is now well on the way to being able to build complex carbon networks to order, just as you intended.

Networking Opportunities
THE FRIEDEL–CRAFTS REACTION

In the final reaction of this part I am going to help you extend your ability to use our toolkit to build a network of carbon atoms. The reaction I talk about here is one of many that I could have chosen and will give you some insight into the way that organic chemists go about building their intricate constructions. It was devised in 1877 by the French chemist Charles Friedel (1832–1899) and the American chemist James Crafts (1839–1917). There are two kinds of Friedel–Crafts reaction: I shall call them Type 1 and Type 2. The latter is more important, but the former is a bit simpler and I shall deal with it first.

> *Pedant's point.* The technical name for Type 1 is 'alkylation' and that of Type 2 is 'acylation'. Type 1 uses an alkyl chloride, Type 2 an acyl chloride.

Theme . . .

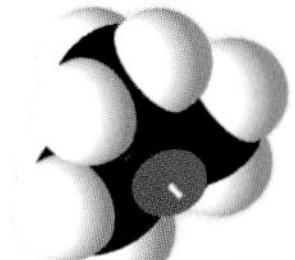

In a Type 1 Friedel–Crafts reaction, the aim is to attach a group of C atoms, such as 1, to a benzene ring or a related molecule. The strategy is to generate a powerful electrophile (Reaction 16), one characteristic of the  group of atoms you want to attach, which will seek out regions of dense electron cloud in the target benzene molecule. The tactics involve taking the group you want to attach in combination with a chlorine

 atom, Cl, as in 2, and then finding another dentist-like compound that will extract the Cl atom as a chloride ion, Cl⁻. That extraction will leave a positively charged hydrocarbon ion hungry for opposite charge and thus able to act as the electrophile.

The Friedel–Crafts procedure uses aluminium chloride, 3, to act as this dentist compound. It gets regenerated in the reaction, so it is present as a catalyst (Reaction 11). When you examine this molecule you see that although its Cl atoms are rich in electrons, the aluminium atom, Al, has a very skimpy share in them and the positive charge of its nucleus shines through. Moreover, the molecule is flat, and there is plenty of room for the Cl atoms to bend away from any incoming

 intruder atom and so make room for its attachment to the Al atom. These features imply that 3 can act as a Lewis acid, an electron-pair acceptor (Reaction 9) and use its attachment to the incoming atom to extract it from its parent molecule.

Let's see what happens when aluminium chloride and the chlorine compound are added to benzene (Figure 22.1). We see the aluminium catalyst molecule bump into the reactant molecule 2. The electron-wispy Al atom of the aluminium molecule latches onto the dense electron cloud

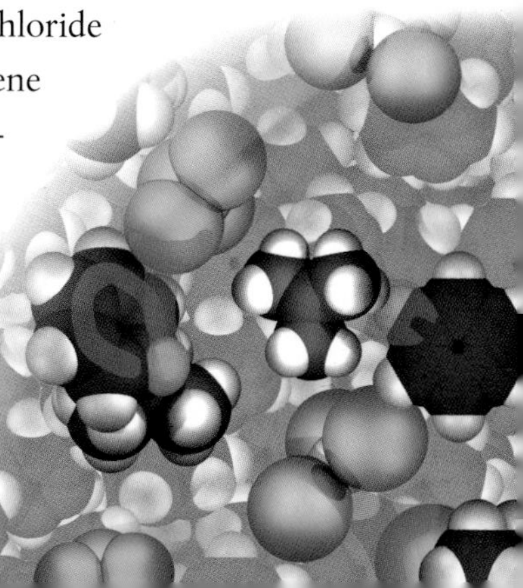

FIG. 22.1 *The attachment of carbon*

around the Cl atom of 2 and draws it out of the molecule as a Cl^- ion. They wander off together into the sunset as an $AlCl_4^-$ ion. The extraction of this atom lets the rest of the molecule relax into a flat shape to relieve the crush of the three CH_3 groups. This relaxation helps the extraction on its way. At this stage we see that we have a positive ion, a potent electrophile.

As we continue to watch we see the ion bump into one of the benzene molecules that surround it. There is plenty of electron cloud on each C atom of the ring, and the ion has no difficulty in pulling electron cloud out towards its central C atom and forming a new carbon–carbon bond, just as you want. You should notice that an H atom is still linked to the ring and so the favourable spread of electron cloud round the ring is disrupted. The molecule has no difficulty in spitting out that disruptive H atom (as a proton), as it may attach to any base. In one instance we see the proton attach to a Cl atom making up an $AlCl_4^-$ ion, with the result that an HCl molecule is formed and the catalyst is free to catalyse a further reaction event elsewhere in the mixture. With the expulsion of the proton, the electron cloud becomes continuous around the ring. The outcome is a benzene ring decorated with a knobbly carbon side chain, just as you wanted.

. . . and Variation

The problem with the Type 1 Friedel–Crafts reaction is that the newly added group of atoms pumps electrons into the benzene ring to which it is now attached. The electron cloud in the ring is thereby intensified and acts as an even brighter beacon than before for the electrophiles lurking shark-like in the mixture. The consequence is that the ring gets attacked again, and a possibly unwanted second group of atoms is added by further electrophilic attack.

The Type 2 reaction is based on a clever strategy designed to prevent that happening. Suppose you were Friedel or Crafts, what would

FIG. 22.2 *Another type of attachment*

you do? You would want to add a group of atoms
to the ring that sucked electron cloud out of it,
making it less appetising to an electrophile and
so avoid further attack. You know from other reactions that an O atom is often a good electron sucker,
so instead of a plain hydrocarbon group you might think of
using a similar molecule but with an O atom somewhere near the C
atom that is destined to form the new carbon–carbon bond. Indeed, a
Type 2 Friedel–Crafts reaction uses not 2 but the oxygen-bearing 4 as
a reagent (I have simplified the molecule for clarity).

We shrink to molecular size and watch what happens during a
Type 2 reaction (Figure 22.2). We may suspect that we will see similar
events with changes of detail because a different reagent is present.
Indeed, we see an aluminium catalyst molecule bump into 4 and do
its normal dentistry: it extracts a Cl atom as a Cl⁻ ion and wanders
off as an $AlCl_4^-$ ion. The extraction leaves a CH_3CO^+ ion in place of
the hydrocarbon ion present in the Type 1 reaction. This ion, with its
positive charge largely centred on the C atom is an electrophile. We

see it seek out a benzene ring and form a carbon–
carbon bond. Just as before, the ring cannot toler-
ate the H atom being still attached, so it spits out a
proton on to a nearby $AlCl_4^-$ ion, so regenerating
the catalyst molecule. We see that the CH_3CO group
is now attached to the ring, which is nearly what you wanted.

I say nearly. There is still present the O atom introduced to ensure
that the reaction stops at this stage by withdrawing electrons from
the benzene ring and inhibiting further electrophilic attack. Perhaps
you want it to remain as part of the architecture. If not, there are various ways of getting rid of it. You saw one way in Reaction 21, where a
phosphorus compound was used to pull an O atom out of a molecule
and replace it with a C atom. Alternatively, you could get rid of the
oxygen by a reduction reaction (Reaction 4) of some kind.

III

MAKING LIGHT WORK

There is another tool that I have delayed introducing until now because it is so potent, more chain saw than hacksaw, for as well as being capable of delicate construction it can also damage and kill.

Light, a bringer of energy, is a potent force for bringing about a chemical reaction. For our purposes it is best to think of light as a stream of particles called 'photons'. Each photon is a packet of energy. Think of red, yellow, and blue blobs of light. The energy of each photon depends on the frequency and hence the colour of the light: photons of low-frequency infrared radiation are packets of relatively small amounts of energy; photons of high-frequency ultraviolet radiation are packets of blistering energy. Photons are like projectiles that plunge into molecules and, if they carry enough energy, shatter bonds, sever atoms from their parent molecules, and blast away electrons. It is time to watch what light can do.

Dark Matter

PHOTOCHROMISM

Let's start with the 'ionization' of an atom, the formation of an ion by the ejection of an electron when the atom is struck by a sufficiently energetic photon. This kind of process is the basis of the operation of early versions of photochromic glasses, which darken when exposed to bright sunlight, specifically in response to the high-energy ultraviolet component of sunlight that is present outdoors. That kind of photochromic glass was made by adding silver and copper nitrates to molten glass. As the glass cools, small crystallites of the salts form. The crystallites are too small to scatter or absorb visible light, so the glass appears transparent.

Now we step into the solid glass and watch what happens when we step outside and ultraviolet photons rain down on us (Figure 23.1 on the next page). We see a photon plunge into the glass and strike a copper ion, Cu^+. The photon has enough energy to expel an electron

FIG. 23.1 *Copper reducing silver*

from the ion, so forming Cu^{2+}. We see the ejected electron wander off through the solid. Almost immediately, however, it is captured by a silver ion, Ag^+, converting it to a silver atom, Ag. (Recall that Ag is the chemical symbol for silver, from the Latin *argentum*.) Sunlight has induced a redox reaction, an electron transfer reaction (Reaction 5). Now, as we continue to watch, several Ag atoms cluster together to give a microscopic dot of silver metal. These little clusters act like tiny shutters to block some of the light passing through the glass, and the image is dimmed.

The clusters of atoms survive for a short time, but inevitably break up and release the additional electron back into the solid. It finds its way back to the strongly attracting double positive charge of a Cu^{2+} ion and attaches to it, so recreating the original Cu^+ ion. However, if you stay outside in the sunlight, ionization and Ag atom formation continue, and your glasses stay dimmed. Only when you come back indoors and the ultraviolet radiation no longer reaches you do the ionization processes cease, and the glass reverts to being fully transparent.

One problem with photochromic glass of this kind is that because the active material is distributed through the glass, thick regions of glass are darkened more than thin regions. Modern sunglasses are typically built from plastic (polycarbonates), and the photochromically active species are organic molecules that are dispersed in a thin film applied uniformly to the surface of the lens.

One photoactive material is based on compounds called 'spiropyrans', **1**, the name reflecting their twisted shape As you can see, in

the resting state of the molecule there are two regions of electron cloud, each spreading over its own territory in different parts of the molecule. Now watch what happens when an energetic

REACTIONS

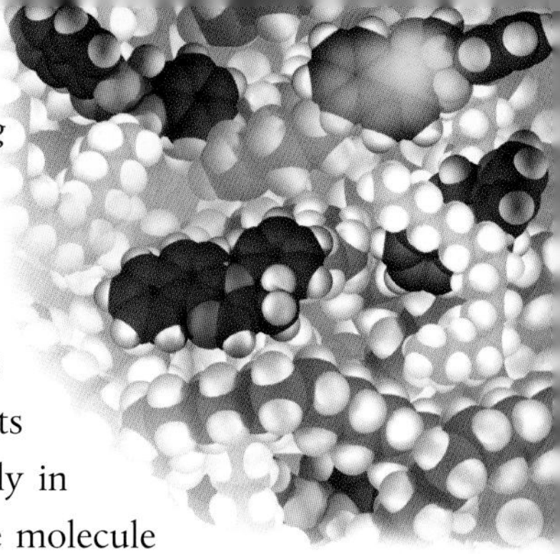

FIG. 23.2 *Untwisting*

ultraviolet photon is absorbed (Figure 23.2). The photon plunges into the electron cloud, like an asteroid splashing into an ocean. It drives some of the cloud out of the region where it acts as glue between two atoms and holds it rigidly in its spiral shape. With that bond weakened, the molecule twists around the surviving link and becomes more or less flat. With all the atoms in the same plane, the electron clouds can merge and a single cloud spreads over the molecule. When electrons are spread out over a wide region they can be excited by the much lower energy photons of visible light. If they absorb relatively smoothly across all frequencies of visible light, then the view is greyed out. If they absorb smoothly only towards the blue and green frequencies, then the image is dimmed but tinted brown.

Irritating Atmospheres

ATMOSPHERIC PHOTOCHEMISTRY

The problem of photochemically generated smog begins inside internal combustion engines, where at the high temperatures within the combustion cylinders and the hot exhaust manifold nitrogen molecules and oxygen molecule combine to form nitric oxide, NO. Almost as soon as it is formed, and when the exhaust gases mingle with the atmosphere, some NO is oxidized to the pungent and chemically pugnacious brown gas nitrogen dioxide, NO_2, 1.

We need to watch what happens when one of these NO_2 molecules is exposed to the energetic ultraviolet photons in sunlight (Figure 24.1 on the next page). We see a photon strike the molecule and cause a convulsive tremor of its electron cloud. In the brief instant that the electron

1

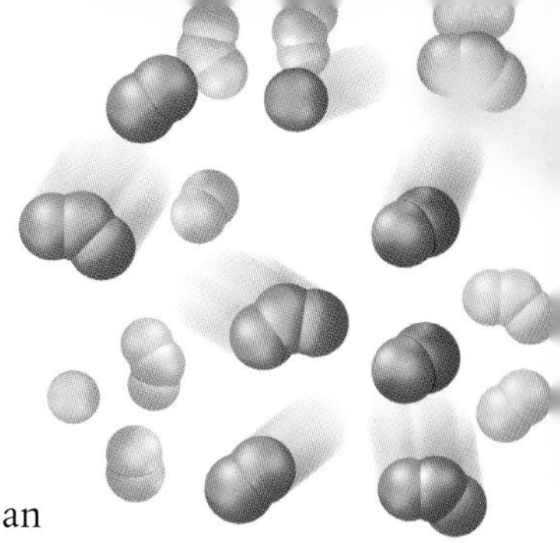

FIG. 24.1 *Ozone formation and destruction*

cloud has swarmed away from one of the bonding regions, an O atom makes its escape, leaving behind an NO molecule. We now continue to watch the liberated O atom. We see it collide with an oxygen molecule, O_2, and stick to it to form ozone, O_3, 2. This ozone is formed near ground level and is an irritant; ozone at stratospheric levels is a benign ultraviolet shield. Now keep your eye on the ozone molecule. In one instance we see it collide with an NO molecule, which plucks off one of ozone's O atoms, forming NO_2 and letting O_3 revert to O_2.

> *Pedant's point.* The formation of O_3 from NO_2 and O_2, the first of these processes, dominates when the concentration of NO_2 is about three times higher than that of NO. The elimination of ground-level destructive, irritating ozone by NO, the second type of process, is dominant when the concentrations of NO_2 and NO are reversed, with NO more abundant than NO_2.

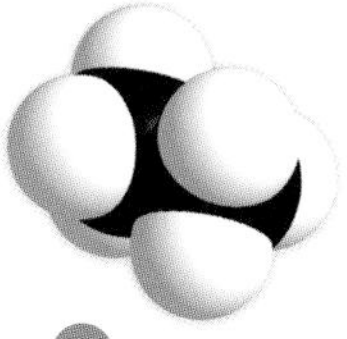

Another fate awaiting NO_2 is for it to react with oxygen and any unburned hydrocarbon fuel and its fragments that have escaped into the atmosphere. We can watch that happening too where the air includes surviving fragments of hydrocarbon fuel molecules (Figure 24.2). A lot of little steps are involved, and they occur at a wide range of rates. Let's suppose that some unburned fuel escapes as ethane molecules, CH_3CH_3, 3. Although ethane is not present in gasoline, a $CH_3CH_2\cdot$ radical (Reaction 12) would have been formed in its combustion and then combined with an H atom in the tumult of reactions going on there.

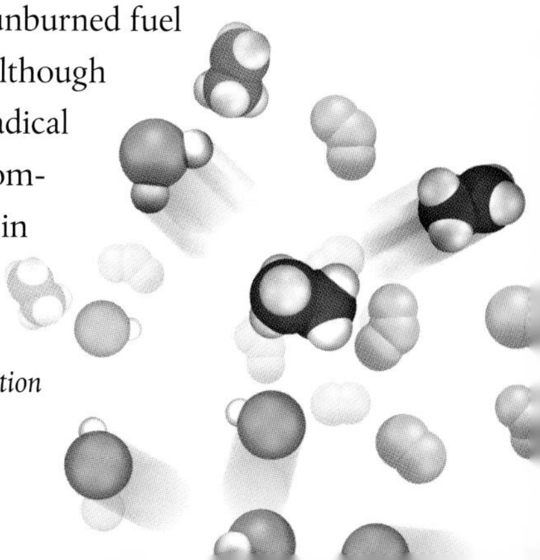

FIG. 24.2 *Hydrocarbon participation*

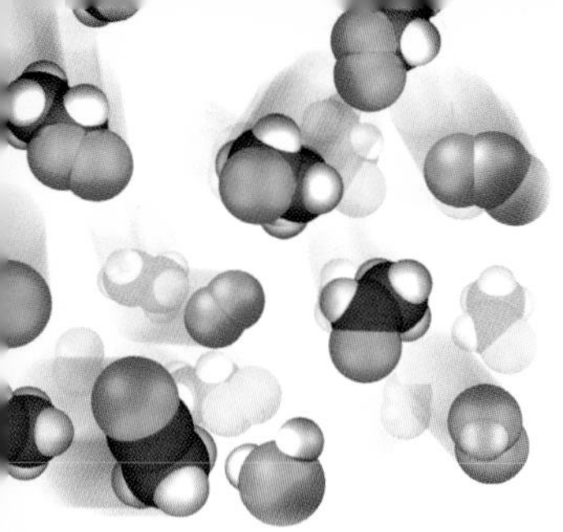

FIG. 24.3 *Radical propagation*

You already know that vicious little O atoms are lurking in the sunlit NO_2-ridden air. We catch sight of one of their venomous acts: in a collision with an H_2O molecule they extract an H atom, so forming two ·OH radicals. The two radicals spring apart, and we see one of them collide with an ethane molecule and scurry off with an H atom, having become H_2O again. The collision has left the ethane molecule as the radical CH_3CH_2·.

We now keep our eyes on the radical, as it is going to grow into a significant component of smog (Figure 24.3). It is surrounded by oxygen molecules and very soon it collides with one and sticks to it. It doesn't take long before this new radical bumps into an NO molecule that is lurking in the polluted air. In the collision, the NO extracts an O atom and flies off as NO_2, leaving the radical with one less O atom. This new radical collides almost immediately with an oxygen molecule, which snips out an H atom and departs as the radical HO_2· to do further damage elsewhere. The molecule left behind, CH_3CHO, has a short life as it is surrounded by radical enemies. In particular the ·OH radical produced in an earlier step is still lurking nearby or skidding in from its formation elsewhere. In the inevitable collision, the ·OH radical extracts an H atom and scurries off as H_2O. That leaves behind yet another radical, CH_3CO.

Like most radicals, this newly formed one is highly reactive, and almost immediately forms a bond to the oxygen molecules that surround and pepper it with collisions (Figure 24.4). The outcome is yet another radical. Lurking nearby we see an NO_2 molecule, which is itself a radical. We see them collide and snap together to form the compound known as PAN, **4** (peroxy-acetyl nitrate). The chain of reactions ceases. PAN is highly soluble in water and moisture and is one of the principal eye irritants of photochemical smog.

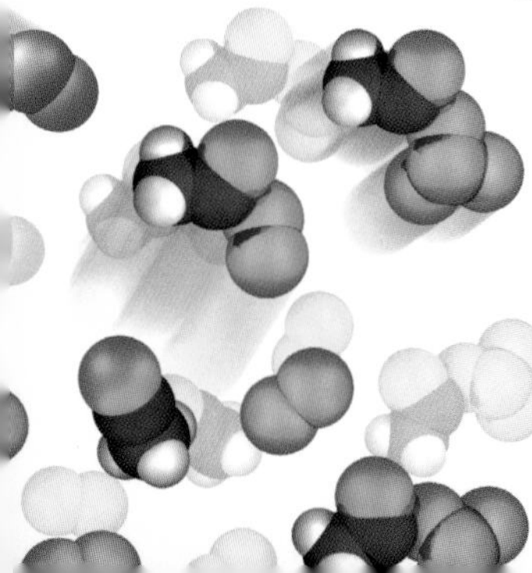

FIG. 24.4 *Irritant formation*

4

25

Seeing the Light
VISION

Now for that extraordinary and wonderful sense, vision. A lot of physics and physiology goes on between the object observed and the focus of its image on the retina of the eye, but the interface of the image with the brain is photochemical. About 57 per cent of the photons that enter the eye reach the retina; the rest are scattered or absorbed by the ocular fluid, the fluid that fills the eye and helps to maintain its shape.

You need to know that the 'rods' and 'cones', the physical receptors distributed over the retina, contain a molecule called retinal, 1, which is anchored to a protein, opsin, to give the aggregate known as rhodopsin. Here the primary act of vision takes place, in which that rhodopsin absorbs a photon. Rhodopsin is the primary receptor for light throughout the animal kingdom, which indicates that vision emerged

1

very early in evolutionary history, no doubt because of its enormous value for survival. Incidentally, a retinal molecule resembles half a carotene molecule, one of the molecules that contribute to the colour of carrots, which is why there is an at least apocryphal connection between eating carrots and improving one's vision.

Let's stand almost literally eye-to-eye and watch what happens when a retinal molecule in your eye absorbs a photon that might have bounced off this page as you read it (Figure 25.1). I have indicated the shape of the opsin molecule by ribbons which show in a general way where its numerous atoms lie. The photon passes through your pupil, negotiates the ocular fluid, and plunges into the retinal hot-spot of rhodopsin. We see it stir up the electron cloud of the long chain of carbon atoms in the tail of the retinal molecule. That stirring briefly loosens the double bond character of the links between the atoms, and the molecule snaps into a new shape with the carbon tail now straight.

The storm in the electron cloud subsides and all the bonds are restored, but now the retinal molecule is trapped in its new shape. Its newly straightened tail results in the molecule pressing against the coils of the big opsin molecule that surrounds it. That change sends a pulse of electric potential through the optical nerve and into the optical cortex, where it is interpreted as a signal and incorporated into the web of events we call 'vision'.

The original resting state of the retinal molecule is restored by a series of chemical events ultimately powered by the metabolism of your ingested food (Reaction 27). I won't go into detail, because re-winding retinal so that it can respond again involves a lot of steps with enzymes. In short, the straightened retinal molecule squirms out of its home in the opsin molecule. Then an enzyme grabs hold of it and bends it back into its original shape. Finally, it is delivered back home, ready to respond again.

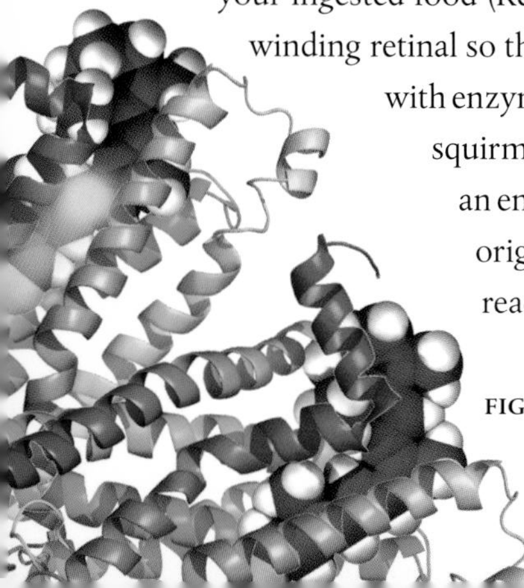

FIG. 25.1 *The primary visual act*

Green Chemistry
PHOTOSYNTHESIS

Each square metre of the Earth receives up to 1 kW of solar radiation, with the exact intensity depending on latitude, season, time of day, and weather. A significant amount of this energy is harnessed by the almost magical process we know as 'photosynthesis' in which water and carbon dioxide are combined to form carbohydrates. Thus, from the air and driven by sunlight, vegetation plucks vegetation. That new vegetation is at the start of the food chain, for its metabolism is used to forge protein and, in our brains, drive imagination. There is probably no more important chemical reaction on Earth.

A large proportion of solar radiation is absorbed by the atmosphere. Ozone and oxygen molecules absorb a lot of ultraviolet radiation, and carbon dioxide and water molecules absorb some of the infrared radiation. As a result, plants, algae, and some species of bacteria have to make do with what gets through and evolved apparatus that captures principally visible radiation. The early forms of these organisms stumbled into a way to use the energy of visible radiation, which arrives in the packets we call photons, to extract hydrogen atoms from water molecules and use them and carbon dioxide to build carbohydrate molecules, which include sugars, cellulose, and starch.

The oxygen left over from splitting up water for its hydrogen went to waste. Most of the oxygen currently in the atmosphere has been generated and is maintained by photosynthesis since Nature first stumbled on the process about 2 billion years ago and thereby caused the first great pollution. That pollution, in Nature's characteristically careless and wholly thoughtless and unplanned way, was to turn out to be to our great advantage.

Photosynthesis begins in the organelle (a component of a cell) known as a 'chloroplast', so you need to poke around inside one if you are to understand what is going on. I shall focus on the light harvesting and the accompanying 'light reactions'. What follows them, the so called 'dark reactions' in which the captured energy is put to use to string CO_2 molecules together into carbohydrates, is controlled in a highly complex way by enzymes. Fascinating as it is, it is far too intricate to make a good story here.

The first thing we notice once we scramble inside one of the many chloroplasts in a cell is an array of radio telescopes. Well, they are not really radio telescopes but have a similar radiation-gathering function. They are the so-called 'light-harvesting complexes', each one consisting of a lot of chlorophyll molecules associated with a large protein molecule. The chlorophyll molecules are the antennae that capture the incoming energy and put us in touch with the Sun (Figure 26.1). Scattered around among them are molecules like carotene (which I mentioned in Reaction 25), which also respond to incoming photons and help to capture their

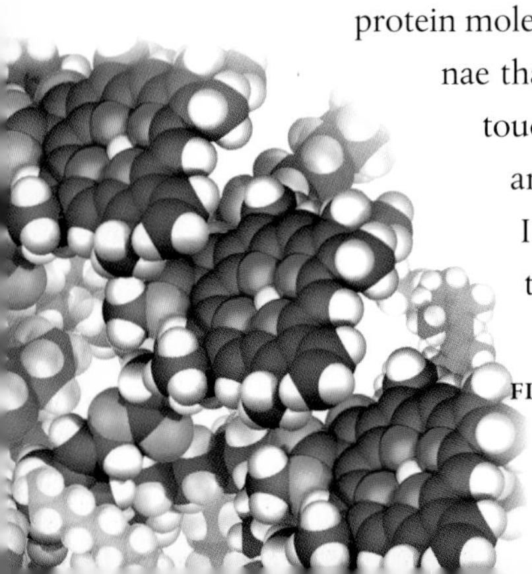

FIG. 26.1 *The antenna array*

energy. Carotene molecules are typically various shades of yellow, and we become aware visually of their presence in leaves when the more fragile chlorophyll molecules decay in the autumn.

Associated with the chlorophyll antennae are factory-like assemblies known as 'reaction centres' where the preliminary work of carbohydrate building goes on by using the power supplied from the chlorophyll on the roof. The efficient little joint assembly of light-harvesting complex and reaction centre is known as a 'photosystem'.

1

A chlorophyll molecule is a big, intricate, and fragile molecule. Let's take a look at it, 1. We see that it is tadpole shaped, with a big flat head and a long hydrocarbon tail. The big flat head has an appearance similar to that of some other important molecules, including the oxygen-carrying haemoglobin of our blood and the myoglobin that stores oxygen in our muscles. Nature, as usual, is being highly economical and, once settled on a viable design, uses it over and over. There is an important atom at the centre of the ring: in chlorophyll it is an atom of magnesium, Mg (in haemoglobin and myoglobin it is iron). This atom sits comfortably in the central cavity, and helps to hold the ring flat and rigid despite the vigorous electronic events going on around it.

We shall watch what happens when a red or blue photon strikes a chlorophyll molecule (Figure 26.2). Green photons don't have the correct energy to be absorbed and just bounce away, which is why vegetation looks green when chlorophyll is still present.

The impacting photon stirs up the electron cloud on a chlorophyll molecule, which briefly becomes the home of the arriving energy. That

FIG. 26.2 *Energy capture*

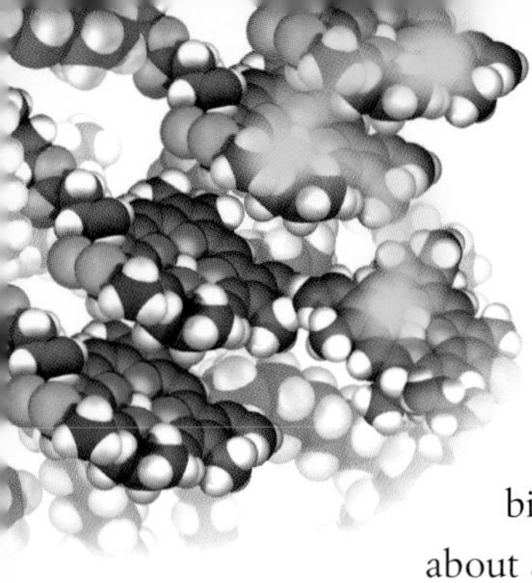

FIG. 26.3 *Energy dumping*

energy is like a hot potato. We see it quickly tossed to a neighbouring chlorophyll molecule, and then to another. In a twinkling of an eye (in more conventional units, in about one ten-billionth of a second) the energy jumps between about a thousand molecules. By passing on rapidly from molecule to molecule, the energy avoids degrading into molecular vibration (heat) or being emitted again as a photon, and is preserved for doing its job of synthesis.

The hot potato finally reaches the factory, the reaction centre itself. It falls onto a pair of chlorophyll molecules and stirs up their electron clouds and then migrates to a closely related molecule, which is essentially a chlorophyll molecule without the central Mg atom (Figure 26.3). The energy now is safely stored and can be used for construction by driving forward a string of proton transfer and redox reactions marshalled and controlled by enzymes.

IV

BUILDING BY DESIGN

You have the toolkit and have assembled the workshop; what about the building? You have seen a number of the techniques that chemists use to build forms of matter, many of which have never existed on Earth before and some conceivably that do not exist anywhere else. But the techniques are merely tools and equipment; they become useful only when set to build something. In this final section I shall lead you through a construction project that uses the tools I have described and, just occasionally, draw on some specialist tools to achieve a particular end. No new principles will be involved, just a slightly different deployment.

I shall describe two kinds of project. The first shows Nature's cunning, illustrating how in the course of just a few billion years she has stumbled unconsciously and blindly into strategies of the greatest subtlety. Then I shall show how ordinary human chemists, who are gradually acquiring great subtlety, achieve construction, in their case with intention and with their eye on a target. In some cases they seek only to add to Earth's store of a particular compound, rendering it cheaper or more abundant so that it can be used more widely. In others they seek to improve on what Nature has already provided and which satisfies her requirements but needs modification for a particular human deployment.

Food for Thought

ENZYME ACTION

Nature makes use of the tools that I have been developing, and does so in the most extraordinary and subtle manner. After all, she has had about four billion years to come up with solutions to problems with which human chemists have striven seriously for only a century or so. Most of the reactions that go on in organisms—including you—are controlled by the proteins called 'enzymes' (a name derived from the Greek words for 'in leaven', as in yeast). Enzymes are biological catalysts (Reaction 11) that are extraordinarily specific and highly effective in their role. One of these complex molecules might serve as the merest foot soldier in the army of reactions going on inside you, with a role such as severing the bond between two specific groups of atoms in a target molecule.

Because their function may be highly specific, enzyme molecules need to be large: they have to recognize the molecule they act on, act on it, then release it so that they can act again. Thus, they have to have several functions built into them. As you will see, enzymes are the ultimate in functional blindness: they feel around in their surroundings in order to identify their substrate, the species they can act on. Life is ultimately blind chemical progress guided by touch.

I am going to introduce you to one particular group of enzymes, the 'proteases', and focus on one example from this group, namely chymotrypsin. A protease is a traitor to its kind: it is a protein that breaks down other proteins. It plays a role in digestion, of course, but its range is much wider. One protease enables a lucky sperm to eat through the cell wall of an egg and ensure its at least temporary immortality. Another facilitates the clotting of blood to terminate possibly fatal bleeding.

Chymotrypsin itself is an enzyme that is secreted from the pancreas into the intestine, and makes an essential contribution to the process of digestion. Its name is derived slightly circuitously from the Greek words for animal fluid, a bodily 'humour', and rubbing, as it was obtained as a fluid by rubbing the pancreas. It is a fair-sized molecule, 1, a long tangled string of pearls (the amino acids in the image I adopted in Reaction 14), consisting of 241 pearls strung together by peptide links, the −CONH− links that I described in Reaction 14. At this stage I am showing you only the general shape of the molecule, not its hundreds of individual atoms.

When it is first made in the pancreas the string is longer, with 245 pearls. It is then sleeper rather than traitor because those additional pearls distort the scrambled chain and inhibit its function. That is an

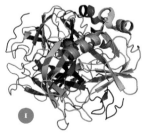

unconsciously wise precaution of Nature, for were the enzyme active it would start to consume the cells that had made it. Perhaps earlier versions did, but natural selection would soon have seen to it that such beings did not survive to propagate themselves.

Once chymotrypsin drips into the small intestine another worker enzyme snips off the four offending pearls and severs the remaining string into three pieces of lengths 13, 97, and 131; the entire assembly can now fold into its active form, which is nearly spherical. The sphere is densely packed with pearls, just as if a real string of pearls were bundled up. If you were to cross the sphere from one side to the other you would encounter up to about ten pearls. The sphere is also sturdy, with a lot of strong sulfur–sulfur bonds to help hold it in shape and withstand the hurly-burly of life in your intestine.

There are, in common with all proteins, large numbers of reasonably strong interactions between the pearls that help to maintain the precision of its shape and to ensure that certain pearls, despite being far apart on the chains are in fact brought close together in the folded molecule. In this form, chymotrypsin is active, and can go about its business of digesting proteins.

I need to say a word about the nature of these interactions, as they will play an important role in what follows. The interaction I have in mind is called a 'hydrogen bond'. By virtue of its tiny size, a hydrogen atom can lie between two other atoms of the appropriate kind, for our purposes either O or N atoms, and stick them together rather like a press-stud joining two pieces of fabric. Hydrogen bonds are much weaker than ordinary chemical bonds but strong enough to contribute substantially to the shape and function of enzyme molecules.

Scissors

Chymotrypsin is a scalpel not an axe. It does not bludgeon and chop every protein it happens to encounter in the slurry of matter passing from your stomach into your intestine but carefully selects its targets. It slices through peptide links that lie next to three neighbouring pearls of a certain type; a common feature of all three is that their

molecules have strikingly flat appendages, as in 2. There is a region near the surface of a chymotrypsin molecule and close to its active site that is similarly shaped and into which these three contiguous flat regions fit snugly. When they lie there, a particular peptide link is brought into the danger zone.

Another molecule, trypsin, is also manufactured and secreted by the pancreas and has a similar function to chymotrypsin, but it recognizes pearls that have side chains that are electrically charged, so jointly the two enzymes can achieve a lot of damage to the proteins you have eaten.

That damage consists of breaking up the string of pearls by hydrolysis of the links between them. You saw exactly the same process in Reaction 17 where I pointed out that these resistant bonds require seriously vigorous conditions, such as boiling for hours in acid. Somehow your body needs to achieve the same ends in the much more benign environment of your gut. In fact, chymotrypsin is remarkably effective at bringing about the reaction, speeding it up compared with ordinary laboratory processes by a factor of about a billion even though the temperature is a mild blood-warm 37°C. I need to show you how it does it. You will see that it brings several tools in chemistry's workshop to bear: it uses nucleophilic attack (Reaction 15), acid catalysis (Reaction 17), and base catalysis (Reaction 18). As in ordinary catalysis (Reaction 11), it finds a way to lower the energy needed to bring about reaction to enable the process to occur at body temperature rather than in boiling acid.

Chainsaws

At this point you accompany me as we lower ourselves cautiously into your postprandial gut, shrink, and watch the action there as a chymotrypsin molecule sidles up to a bit of today's lunch in the form

FIG. 27.1 *Serine at the active site*

of a protein molecule. From now on I need to acknowledge the pearl's individual chemical personalities and call them by their actual names. You already know that they are actually fragments of amino acids. They are not whole amino acids because some of their atoms have been used up to make the peptide links between them. What remains are called 'residues'. That is, each pearl is an 'amino acid residue', more formally a 'peptide residue', with a chemical personality determined by the atoms that remain after the links to its neighbours are formed. Each residue is identified by an abbreviation of the name of its parent amino acid and its numerical location counted along the chain from one end.

We focus on the heart of the active site and in particular a serine residue (Ser-195) that lies there (Figure 27.1). This molecule wields the scalpel, but it does not act alone. It is aided in its slicing by two others, a histidine residue (His-57) and an aspartic acid residue (Asp-102) that although far apart on the linear chain are brought nearby when the protein has folded into its final shape. I shall simplify the appearance of the eaten protein by shaving off the side groups that enable it to be recognized, 3.

We see that the histidine residue has formed a hydrogen bond to the OH group in the serine residue. In doing so, the H atom of that group is pulled away a little from its neighbouring O atom. That pulling leaves the latter with a greater share of the electron cloud forming the oxygen–hydrogen bond. In other words, the electron cloud on the O atom has become more dense, given it a partial negative charge, and turned the O atom into a serious nucleophile (Reaction 15).

We should also take note of the aspartic acid residue lurking near the histidine residue (Figure 27.2). When we look carefully we see that it has

FIG. 27.2 *Aspartic acid assisting*

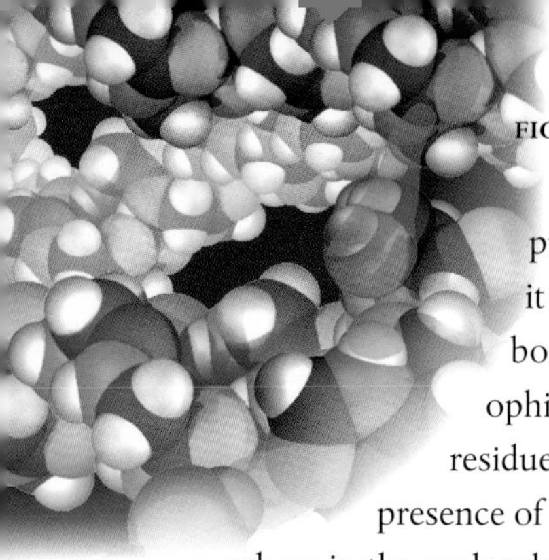

FIG. 27.3 *Serine launches its attack*

pulled on the histidine residue and orientated it into a better position to form the hydrogen bond to serine that. That means that the nucleophilic character of the O atom on the serine residue that I have just described is enhanced by the presence of the aspartic acid residue. Serine residues elsewhere in the molecule are not affected in this way and are so feeble as nucleophiles that they are chemically innocuous; only the serine buried deep down in the active site is activated in this way.

Now watch what happens when the serine residue in the active site finds a peptide bond snuggled unwarily down beside it (Figure 27.3). It bites. Its electron-rich nucleophilic O atom senses the partial positive charge of the C atom of the CO group in the –CONH– peptide link. The electron cloud reaches out to the C atom and a carbon–oxygen bond forms. In the face of this invasion, the electrons cloud of the CO group partially retreats on to the O atom to make room for the incoming electrons. I need to mention a point that will turn out to be significant later: at this stage the atoms are arranged at the corners of a tetrahedron around the central C atom. Just remember that point for now.

Now we see the electron cloud on the O atom of the CO group flood back. There is another retreat in the face of this advance (Figure 27.4). The electron cloud that formed the carbon–nitrogen bond withdraws into the comfort zone of the N atom, and the proton that began life attached to the O atom of the serine residue accompanies them. We see that the strand of protein molecule that ends at the N atom is no longer bound to anything. We watch it float off into the surroundings, where it will make another encounter with a chymotrypsin or trypsin molecule and be broken up still further.

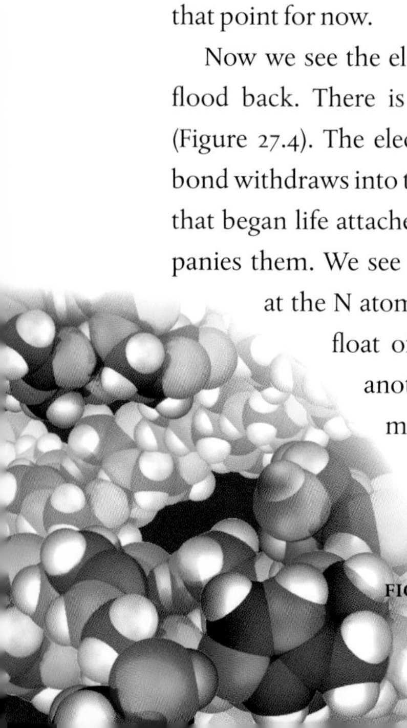

FIG. 27.4 *Scission is complete*

FIG. 27.5 *Base catalysis takes over*

Eating water

Let's continue to watch, because there is still plenty to do. We see that the location vacated by the disappearing part of the protein molecule is immediately filled by one of the water molecules that are abundant in your gut. Now the histidine residue and its collaborator, the aspartic acid residue, can act again, this time focusing on H_2O and preparing it to be their agent.

The potent N atom of the histidine residue pulls on one of the H atoms of the water molecule, and just as the serine residue did before, the O atom of the H_2O molecule acquires an enhanced negative charge (Figure 27.5). The histidine residue goes on pulling and we see the H_2O molecule split into a proton that attaches completely to the seductive N atom and an OH^- ion. The OH^- ion is a powerful nucleophile that immediately attacks the partial positive charge on the C atom of the CO group and forms yet another carbon–oxygen bond. We see the usual electron rearrangement take place. As a result the remaining part of the original protein molecule becomes free and we see it float away to meet its fate elsewhere.

As the process of degradation occurs, the amino acids and fragments of the chain that are released are able to migrate to other regions of the body, where they will be used to construct other proteins, some enzymes and some structural.

Fast food

There is another feature, another contribution to the great acceleration of the reaction other than there being both acid and base catalysis at work in the same microenvironment of the active site. I remarked that in the laboratory boiling acid or alkali is needed to sever a peptide link but in the body the temperature is very much lower. The explanation can be found in yet another subtle aspect of the structure of chymotrypsin. To understand this point, you need to recall from Reaction 11 that reactions go faster if the energy barrier between reactants and products is lowered in some way.

Remember that I drew attention to the tetrahedral arrangement of atoms around a C atom at an intermediate stage of the first nucleophilic attack. If you look very closely at the structure of the enzyme, you will see that a glycine molecule (Gly-193) and our serine molecule (the same Ser-195) can both form hydrogen bonds to a tetrahedral arrangement of atoms in their vicinity. Therefore, because the energy of the tetrahedral arrangement is greatly lowered by this interaction, a much lower temperature can succeed in reaching that arrangement than in the absence of that hydrogen bonding. As a result, the reaction proceeds at a much lower temperature than in the absence of that stabilization. Your gut need not be at boiling point for you to eat successfully.

Grand Designs
SYNTHESIS

In this final reaction I am going to show you in the broadest of outlines how chemists build the equivalent of a cathedral. That is, how they synthesize a complicated molecule from scratch. The aim of a synthesis is to take a reasonably readily available laboratory chemical and process it—add bits on, take things off, close rings of atoms, open rings, build flying buttresses, and so on—until the target compound has been made.

You could take the view that you should really start from absolute scratch, from the elements themselves, typically hydrogen, carbon, nitrogen, and oxygen, and build the molecule from those. However, that would be a waste of time and not crucial to the demonstration of the synthetic route because it is possible to argue that there are already plenty of methods for synthesizing the simple starting materials from scratch, and the real challenge is to build the intricate molecule. That is rather like accepting that a contractor can supply windows, bricks, and beams when constructing a real house and that it isn't necessary to go all the way back to the sand, clay, and iron ore from which they are made to demonstrate that the house can be built literally from the ground up. Of course, the starting materials in a modern chemical synthesis might seem a bit recondite, but be assured that they are reasonably acceptable and purchasable from suppliers of laboratory reagents or easily made from what they do supply.

Now for the particular cathedral on which I intend to focus. That scourge of humanity, malaria ('bad air'), was introduced into the New World in the fifteenth century and soon wrought the havoc that had for long, and still, afflicts millions. The natives there found that an extract of the bark of the *quina-quina tree*, in due course to be classified as *Cinchona Officinalis*, was an effective cure, in particular having saved the life of the Countess of Cinchona. In due course the active component, quinine, was identified and extracted.

Such is the impact of malaria that desperate attempts were made during the nineteenth century to synthesize it from materials present in coal tar, but without success. William Perkin's erroneous approach and his consequent synthesis of the dye mauveine led, however, to another unanticipated and in due course huge success: the foundation of the chemical industry.

Additional pressure to find new sources arose during World War 2, when the cinchona plantations of Indonesia were no longer accessible. A synthetic route was finally devised in 1944 by William von Eggers Doering and Robert Woodward at Harvard. Improved approaches were devised by Milan Uskokovic's team at Hoffmann–La Roche in the USA (in 1970) and Gilbert Stork at Columbia University (in 2001), but none of these intricate and therefore uneconomical syntheses has replaced the natural source. Perhaps the reactions you have encountered in the preceding sections will inspire you to find a simple, economical procedure that, in common with so many other chemical

FIG. 28.1 *Starting materials and the final product*

discoveries, will save the lives of millions.

There are a lot of clever steps in the Doering–Woodward synthesis and I shall do no more than outline their procedure. Each step makes use of one of the tools I have described in earlier sections, but I will leave them implied rather than displayed and described explicitly. In a slight departure from the type of illustrations I have used so far, I will show you the growing components of the final molecule, 1. The two starting materials that you, like Doering and Woodward, will use as a foundation are shown in Figure 28.1. You can see the motif of the former in the quinine molecule, but there doesn't seem to be much sign of the latter. Indeed most of the work of construction involves elaborate manipulations of the second

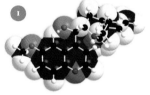

compound, warping it into the complicated structure shown in 1. Then, after the necessary warping, the result is pinned onto the other molecule to make something that is then finally turned into quinine.

It's a footnote to history that Doering and Woodward knew, or thought they knew, that someone else, Paul Rabe, had already shown in the early years of the twentieth century how to achieve the final step, so they thought it wasn't necessary to go that far themselves. Controversy raged for a long time about whether the earlier work was valid: if it wasn't, then Doering and Woodward's synthesis wouldn't have been complete. In the end, as recently as 2008, it was shown that the original work done nearly a century earlier was indeed valid.

Your first step is to start building the right-hand side of the quinine molecule. You first set about building a second ring attached to the benzene ring of the starting material. That involves attaching a group of atoms to the ring and then pinning its far end to a neighbouring point on the ring (Figure 28.2).

FIG. 28.2 *Ring formation*

FIG. 28.3 *Breaking a ring open*

We see the chain of atoms threshing around and then suddenly making a grab for the only C atom on the benzene ring within its reach. It forms a bond, so completing a six-membered ring.

Now you need to grow the side chain on the site next to the OH group and then hack open the benzene ring (Figure 28.3). To do so, you will use a clever version of the elimination reaction (Reaction 20) in combination with a precipitation reaction to get rid of unwanted waste material (Reaction 1: I mentioned that this boring kind of reaction would finally come into its own at some point).

Now you need to join the two fragments of the molecule together by making a new carbon–carbon bond. This crucial step makes use of a reaction devised by yet another German chemist, Rainer Claisen (1851–1930), during the golden age of German chemistry, in about 1881. We watch the two large molecular hulks approach each other in the presence of OH$^-$ ions (Figure 28.4).

There is now only a little bit of tidying up that remains for you to do with a few nips and tucks of atoms here and there, including building a little molecular bridge across the right-hand side of the molecule. You have arrived at your extraordinary destination, quinine. You are, perhaps, not a little chagrined to realize that *Cinchona Officinalis* brainlessly achieves the same result in the quiet unconscious contemplation of its cells without all the paraphernalia you have needed to use, and using a much more sophisticated step under the influence of enzymes. But that's Nature. On the other hand, it is also quite remarkable that chemists have discovered how to emulate her, and using the skills built up over a couple of hundred years to discover how to bend atoms to their will and build from them the equivalent of astonishing cathedrals.

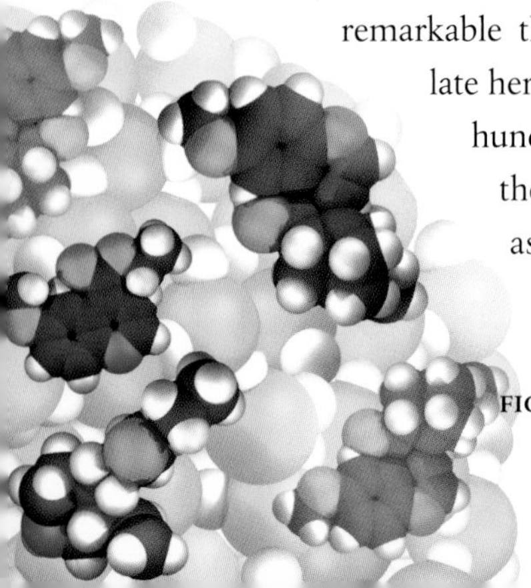

FIG. 28.4 *Joining the parts*

V

ECONOMIZING

There are a lot of tools accumulated in the initial toolkit of Part I, a dozen in all. Some of them are just slightly different versions of others, just as a mallet is a type of hammer. Thus, combustion is an oxidation, reduction is a component of redox reactions, and electrolysis, corrosion, and the generation of electricity are all aspects of redox reactions. Complex substitution is another type of Lewis acid–base reaction. Catalysis makes use of all kinds of the basic tools, and is more like a lubricant than a tool. When you stand back, it might seem that there are just five basic types of reaction: precipitation, proton transfer, electron transfer, Lewis acid–base, and radical recombination. You used them all in the work of construction that followed their assembly.

But can those five types be whittled down to four? In Reaction 9 I argued that a precipitation reaction can be regarded as a Lewis acid–

base reaction, so in fact the five is really four. In the same section I also showed that proton transfer reactions are in fact special cases of Lewis acid–base reactions, in which a proton is simply a special case of a Lewis acid migrating from one complex to another. So that four become three. At this stage it seems that all the reactions I have discussed are electron transfer, electron-pair sharing (Lewis acid–base), or electron sharing (radical combination). Notice how, by a slight shift of nomenclature, I have shifted attention entirely to the behaviour of electrons: their transfer, their sharing, and the sharing of pairs.

Is it possible to go further? Is it conceivable that your toolbox contains only a single tool and that all chemical reactions are aspects of a single process? That would really be something. Well, it is something, for it is possible to identify a single common feature and, with the definition suitably extended, to regard every reaction as an acid–base reaction.

In the same spirit of generalization that characterized Lewis's approach, let's take an extraordinarily bold step and define a base as follows: a base is an electron. In a similarly bold way, let's define an acid as the absence of an electron in an atom or molecule's electron cloud. I shall call that absence a hole, so the proposal is that an acid is a hole and a base is an electron. You can't get more fundamental than that! These definitions are closely related to the more elaborate proposals made by the Russian chemist M. Usanovich in 1938 and which I warmed over elsewhere some time ago.[1]

Let's explore whether all reaction types are captured by these definitions.

Radical recombination reactions plainly are: each radical provides an unpaired electron and, simultaneously, the absence of an electron (a hole) in an electron cloud. Thus, the electron of each radical enters the hole of the other and they jointly form an electron-pair bond.

1   In my *Atoms, Electrons, and Change*, Scientific American Library (1991).

Radical reactions are indeed acid–base reactions (with our new definition of acids and bases).

Lewis acid–base behaviour is a simple extension of this idea, with one species providing both electrons and the other species providing a pair of holes (the absence of two electrons in a cloud), so when complex formation occurs, both the former electrons plug the double hole and form a bond. Lewis acid–base reactions are indeed acid–base reactions (with our new definition of acids and bases).

We have seen that the fundamental criterion of a reaction being redox is that electron transfer must have taken place. That electron comes from one species, and if it is to find a home on the second species, then that species must be able to accommodate it, and must therefore possess a hole. Redox reactions are indeed acid–base reactions (with our new definition of acids and bases).

We can conclude that every reaction is indeed the manifestation of a single event: an electron occupies a hole. Every reaction type is an acid–base reaction.

It might seem rather wonderful to have reduced every chemical event to a single process, but with extreme generality there often comes uselessness. That is the case here. The advantage of the conventional classification into a variety of types is that each reaction has a personality: proton transfer reactions are quite different in practice from redox reactions, and it is not in the least useful, except in some kind of philosophical sense, to lump together redox reactions with, for instance, precipitation reactions any more than it is to classify elephants with dandelions even though both are 'living things'.

That having been said, I would like you to take away a single message. All the wonderful world around us, its materials and its activities, are brought about by conjuring about 100 elements (and far fewer for most important entities and activities) in a tiny number of ways. That potent economy is one of the true glories of Nature.

GLOSSARY

Acid	A proton donor (see *Lewis acid*)
Addition	The attachment of an atom or group of atoms to a molecule
Alcohol	An organic compound of formula R–OH
Alkali	A water-soluble base
Alkane	A hydrocarbon with no multiple bonds.
Alkyl group	A hydrocarbon group derived from an alkane
Alkylation	The attachment of an alkyl group to a molecule
Amide	An organic compound formed from a carboxylic acid and an amine; a compound of formula R–CO–NHR'
Amine	An organic compound of formula R–NH_2
Amino acid	An organic compound of formula $NH_2CHRCOOH$
Anion	A negatively charged atom or group of atoms
Anode	The electrode at which oxidation occurs
Atom	The smallest particle of an element
Base	A proton acceptor (see Lewis base)
Bond	A shared pair of electrons lying between two atoms

Bromonium ion	An organic cation with a positive charge on a Br atom
Carbonium ion	Synonym of carbocation
Carbohydrate	An organic compound of typical formula $(CH_2O)_n$
Carboxylate ion	An anion formed by loss of a proton from a carboxylic acid, $R-CO_2^-$
Carboxylic acid	An organic acid of formula $R-COOH$
Catalysis	The acceleration of a chemical reaction by a species that undergoes no net chemical change
Cathode	The electrode at which reduction occurs
Cation	A positively charged atom or group of atoms
Combustion	Burning in oxygen
Complex	The union of a Lewis acid with a Lewis base
Condensation	The linking of two molecules accompanied by the expulsion of a small molecule (typically water)
Corrosion	The unwanted oxidation of a metal
Dimer	The union of two molecules
Double bond	A link between atoms formed by two shared pairs of electrons
Electric current	A flow of electrons
Electrochemistry	The study of the relation between electricity and chemical reactions
Electrode	A metallic conductor where electrons enter or leave a solution
Electrolysis	To achieve a chemical reaction by passing an electric current
Electromagnetic radiation	Rays of oscillating electric and magnetic fields (as in light)

Electron	The negatively charged subatomic particle that typically surrounds a nucleus
Electrophile	A species that is attracted to electron-dense (negative) regions
Elimination	The removal of atoms from neighbouring sites on a molecule and the consequent formation of a double bond there
Entropy	A measure of molecular disorder
Enzyme	A protein that acts as a catalyst
Equilibrium	A condition in which forward and reverse processes are taking place at the same rate
Ester	A compound formed from a carboxylic acid and an alcohol; a compound of formula $R–CO–OR'$
Excitation	Raising to a state of higher energy
Exothermic	A process that releases energy as heat
Free radical	See *Radical*
Heterogenenous	In different physical states
Homogeneous	In the same physical state
Hydration	The addition of water; surrounded by water molecules
Hydrocarbon	A compound of carbon and hydrogen
Hydrolysis	Breaking down by the addition of the elements of water (H and OH)
Hydronium ion	The ion H_3O^+
Hydroxide ion	The ion OH^-
Intermediate	See *Reaction intermediate*
Infrared radiation	Long wavelength, low frequency electromagnetic radiation (below red)
Ion	An electrically charged atom or group of atoms (see *Cation* and *Anion*)

Ionization	The formation of an ion by the ejection of an electron
Isomer	Structural variants of a compound
Isomerization	The conversion of one isomer to another
Lewis acid	An electron pair acceptor
Lewis base	An electron pair donor
Ligand	A group of atoms bound to a central metal atom
Lone pair	A pair of electrons not involved directly in bond formation
Molecule	The smallest particle of a compound; a specific group of bonded atoms
Monomer	The unit from which a polymer is built
Neutralization	The reaction of an acid with a base to form a salt
Nucleophile	A species that is attracted to electron-poor (positive) regions
Orbital	The region of space occupied by an electron in an atom, molecule, or ion
Oxidation	The removal of electrons from a species; reaction with oxygen
Peptide link	The group of atoms –CONH– linking amino acids in proteins
Peptide residue	See *Residue*
Photochemistry	The study of chemical reactions caused by light
Photochromism	The generation of colour by the action of light
Photoisomerization	Isomerization caused by light
Photosynthesis	The formation of carbohydrates by the action of light

Photon	A particle of electromagnetic radiation
Pi bond (π bond)	The second component of a double or triple bond formed by the side-by-side overlap of orbitals on neighbouring atoms
Polymer	A molecule formed my joining together many small molecules
Precipitation	Coming out of solution in a finely divided form
Product	The material produced by a chemical reaction
Protein	A complex compound built from amino acids
Protease	An enzyme that breaks down other proteins
Proton	The nucleus of a hydrogen atom
Radical	A species with at least one unpaired electron
Reactant	The starting material in a specified chemical reaction
Reaction intermediate	A species other than the reactants and products that is proposed to be involved in a reaction mechanism
Reagent	A substance used as a reactant in a variety of chemical reactions
Redox reaction	A reaction involving oxidation of one species and reduction of another; an electron transfer reaction
Reduction	The addition of electrons to a species; reaction with hydrogen
Residue	The component of a polypeptide chain derived from an amino acid
Salt	An ionic compound formed by a neutralization reaction
Single bond	A link between atoms formed by one shared pair of electrons

Solute	A dissolved substance
Solvent	The medium in which a solute is dissolved
Species	Used here to denote an atom, molecule, or ion
Substitution	The replacement of one atom or group of atoms by another
Synthesis	The construction of molecules
Triple bond	A link between atoms formed by three shared pairs of electrons
Ultraviolet radiation	Short wavelength, high frequency electro-magnetic radiation (beyond violet)

The elements referred to in this account and their locations in the Periodic Table

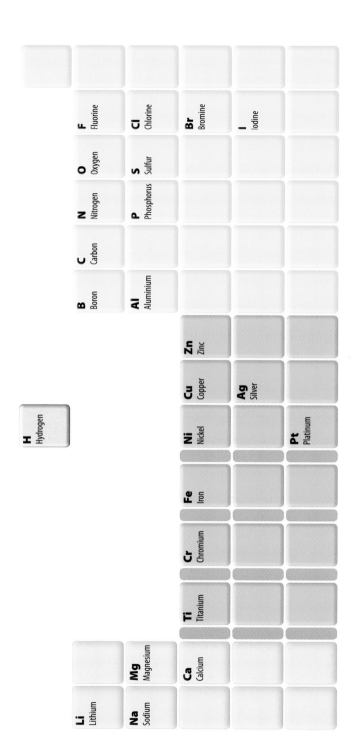

INDEX

INDEX

INDEX